Strickland Landis Kneass

Practice and Theory of the Injector

Strickland Landis Kneass

Practice and Theory of the Injector

ISBN/EAN: 9783337106317

Printed in Europe, USA, Canada, Australia, Japan

Cover: Foto ©ninafisch / pixelio.de

More available books at **www.hansebooks.com**

PRACTICE AND THEORY

OF THE

INJECTOR.

BY

STRICKLAND L. KNEASS, C.E.,

MEMBER OF THE AMERICAN SOCIETY OF MECHANICAL ENGINEERS, FRANKLIN
INSTITUTE, ENGINEERS' CLUB OF PHILADELPHIA.

SECOND EDITION.
REVISED AND ENLARGED.
FIRST THOUSAND.

NEW YORK:
JOHN WILEY & SONS.
LONDON: CHAPMAN & HALL, Limited.
1898.

PREFACE.

Although much has been written concerning the theory and the action of the Injector, there have been few books published since the appearance of Giffard's own pamphlet in 1860, which have been based directly upon experimental research.

It has been the object in the following pages to present solutions of some of the more interesting problems, with illustrations drawn from practical tests, and to describe in detail the function of the different parts. To the professional engineer and to the student, theoretical discussion of the Injector is a tempting field, because of the beauty of its underlying principle and by reason of the numerous associated problems of fluid motion ; the analysis of this part of the subject, based upon carefully conducted laboratory tests, has been fully treated, and complex formulæ have been avoided in the mathematical discussion.

CONTENTS.

CHAPTER I.
EARLY HISTORY 1

CHAPTER II.
DEVELOPMENT OF THE PRINCIPLE 8

CHAPTER III.
DEFINITION OF TERMS—DESCRIPTION OF THE IMPORTANT PARTS OF THE INJECTOR 25

CHAPTER IV.
THE DELIVERY TUBE. 30

CHAPTER V.
THE COMBINING TUBE 44

CHAPTER VI.
THE STEAM NOZZLE 53

CHAPTER VII.
THE ACTION OF THE INJECTOR 67

CHAPTER VIII.
APPLICATION OF THE INJECTOR—AMERICAN AND FOREIGN PRACTICE 89

CHAPTER IX.
DETERMINATION OF SIZE—TESTS 125

CHAPTER X.
REQUIREMENTS OF MODERN RAILROAD PRACTICE—REPAIRS—METHODS OF FEEDING LOCOMOTIVE BOILER 149

INDEX 159

THE GIFFARD INJECTOR.

CHAPTER I.

EARLY HISTORY.

To HENRI JACQUES GIFFARD, an eminent French mathematician and engineer, belongs the honor of having invented the simplest apparatus for feeding boilers that has ever been devised, utilizing in a novel and ingenious way the latent power of a discharging jet of steam.

From the time of his graduation from *L' Ecole Centrale* in 1849, Giffard had directed his energies to the study of aeronautics and had spent much time in developing a light steam motor for propelling balloons; it is, therefore, not strange that he should also have attempted to devise a compact and convenient substitute for the steam pumps then in use. Already a number of patents had been granted him for the application of the steam engine to aerial navigation and for other correlated inventions when, on May 8, 1858, letters patent were issued for *L' Injecteur Automoteur*. His early technical education and wonderful ingenuity well fitted him for breaking away from the old beaten paths and starting out on a new line of discovery; and in view of the originality of his work he fully deserved the unqualified praise accorded him by his contemporaries.

Upon purely theoretical grounds the method by which he proposed to force a continuous stream of water into the boiler appeared to be entirely feasible and would, if practicable, possess many advantages over the intermittent systems. The difficulty seemed to lie in fulfilling the peculiar conditions required for the condensation of the steam and the subsequent reduction of the velocity of the moving mass. Giffard carefully considered the various phases of the question and made a working drawing embodying his ideas. A model was made by M. Flaud & Cie., of Paris, who found, however, considerable difficulty in forming the tubes in the peculiar shapes required. But in the shape and proportions of the nozzles lay the element of success, and the first instrument constructed entirely fulfilled the expectation of the designer.

There have been few other inventions in which the underlying principles have been so thoroughly worked out by the original inventor. Giffard seems to have made a very complete survey of the possibilities of the Injector prior to placing it before the public, and in his patent specification, describes a number of improvements that have since been made. In 1860 he published a small brochure entitled "A Theoretical and Practical Paper on the Self-acting Injector," in which he says: "Of all the necessary accessories of a Steam Engine, perhaps the most important is the one used for feeding water to the boiler; upon its proper working depends not only the regular running of the engine, but the safety, the very existence of those who approach the boiler; . . . . nevertheless, by a kind of fatality, the apparatus employed up to the present time for feeding is, of all others, that which leaves most to be desired." After reviewing the disadvantages of the various methods in use, he continues, "It is important, therefore, to create a new method, free from the imperfection and inconvenience pointed out," and modestly adds, "Such is, it appears to me, the result obtained by the apparatus to which I have given the name of *Injector*, because it produces a veritable continuous injection. Its mode of action, extraordinary in appearance, contrary to that which we are in the habit of seeing or supposing, is explained by the simplest laws of mechanics and has been foreseen and calculated in advance." He describes his invention in detail and explains very fully the best proportions for its various parts, and also the mechanical theory, substantially as advanced by him in 1850, eight years before the construction of his experimental Injector.

And yet, in common with all new inventions and radical improvements, great difficulty was at first experienced in obtaining a fair trial of its merits, and in many cases the exaggerated claims of its friends interfered as much with its early adoption as the openly expressed criticism of its enemies. The great advantages of the new method were appreciated, however, by the *Academie des Sciences* of France, who awarded Giffard the Grand Mechanical Prize for 1859. This was all the more complimentary as it was entirely unsolicited. Prominent engineers presented before the principal scientific societies analytical demonstrations of the theory of the injector and allayed to a great extent the suspicion in the popular mind that the inventor was encroaching dangerously near the claim for perpetual motion. Combes, Bougere, Reech, Villiers, Zuber and Pochet are among the most prominent scientists who made a special study of the subject, and the demonstration of Pochet is still frequently used in modern text books.

It must not be supposed that Giffard was alone in his efforts to utilize the power of a discharging jet. For exhausting and pumping purposes we have record that a crude ejecting apparatus had been used as early as 1570 by Vitrio and Philebert de Lorme. But the first device that bears any similarity to the principle of the Injector was patented August 15, 1818, by Mannoury de Dectot, who describes "sundry motors or means for employing the power of fire, of steam, of air, etc., to start the movement of machines." He applied his invention for raising water and for propelling boats by utilizing the expansion and condensation of steam in connection with jets of water.

Ravard followed in 1840 with improved forms, but the greatest advance was made by Bourdon, the celebrated inventor of the metallic steam gauge, who approached very near the results obtained by Giffard. Two patents were issued to Bourdon, one in 1848 and one in 1857, but it is to the latter that special reference will be made. This contained numerous combinations of convergent and divergent tubes for transforming the energy of a moving jet, or for discharging large or small quantities of liquids or gases. The similarity of the form of the apparatus to that of Giffard was so marked that the question of priority at once arose and was exhaustively discussed by the "*Société des Ingénieur Civils.*" It was shown that Giffard was wholly unaware of the last improvement of Bourdon when he applied for his patent, and as he had publicly presented the theory of his invention nearly seven years in advance of Bourdon, full credit was given him for the conception of the Injector and originality in the application of the principle.

The introduction of the injector into England by Sharp, Stewart & Co., of Manchester, is thus described by one thoroughly familiar with its history and to whom its early success in that country was in great measure due:

"In the autumn of 1859 when our representative in Paris sent over to me a No. 4 Injector as a curiosity and engineering anomaly, he told me simply what it did, but gave no instructions for fixing or working. At about the same time the Paris representative of Messrs. Robert Stephenson & Co., Newcastle, sent over to them a similar Injector. I set to work at once, and by good luck coupled up the correct pipes to their proper flanges, but was a great deal bothered what to do with the overflow flange. After a few nights' work I got my Injector fixed and got up steam, and to some extent began clumsily experimenting as the pressure rose to 60 pounds, the full working pressure of the boiler. I had the Injector fixed over a tank fed by a ball tap and closed by the boiler. I turned steam on and was staggered by the rush of water into the tank from the overflow pipe, and thought something was wrong. However, I continued to turn the steam spindle, and the escape from the overflow sensibly diminished until I could turn no further. In the mean time the ball tap started running furiously into the tank, showing me that water was going somewhere and I knew it could go nowhere else but into the boiler. I then began to operate with the four thread screw at the side, and found that it adjusted the water supply, and succeeded in getting the overflow "dry." I then opened the peep-holes opposite the space between the combining and the receiving nozzles, and saw the white stream passing from one to the other on its way to the boiler. I then ceased operations, and had a pipe of tobacco, and let some water out of the blow-off cock; then I tied a piece of spun yarn round the glass water gauge to prepare for another start, and shortly after, the senior partner came round for a stroll and found me operating. I stopped it, started it, and regulated it so much to his satisfaction that within one week the monopoly of its manufacture in England was secured by the firm. Unfortunately for Stephenson & Co., they coupled their sample Injector up incorrectly, and it would not work."

For stationary service the Injector did not at first become popular; possibly on account of the mystery that seemed to surround its working, and the general skepticism as to its practical wearing powers. Some of the contributions and queries published in the engineering papers of the day, are very amusing, and a certain writer in one of the most prominent weeklies proves most conclusively to his own and probably to some of his readers' satisfaction, that the new method of feeding boilers was an absolute impossibility. The injector was, however, adopted in many places and continued to give satisfaction. In the first trip of the "Great Eastern" Injectors were used in place of pumps, but for some reason not explained, they were subsequently removed; this may have been owing to the temperature of the feed water being too warm for efficient service, as this was the weak point of the first injectors constructed.

The first injector applied to a locomotive in England was by Mr. J. Cross, Superintendent of the St. Helens Railway. It was successful from the start, although not large enough for the purpose and therefore a No. 8 was substituted, which proved to be entirely satisfactory.

The English railroads opened a wide field for the Injector; upon most of the locomotives, the earliest feeding pumps were worked by hand, but afterwards coupled to a special eccentric or to the crosshead. Stretton, in his recent work on the Locomotive, says that it was a common occurrence for engines with a single pair of driving wheels, to stand on well greased rails with tender brakes fast locked and drivers revolving, in order to fill the boiler full of water. But even though the old methods were very crude, engineers in England were much prejudiced against any change, even for the better. By way of illustration the following incident may be given of a successful attempt to convince an obstinate engineer against his will of the advantages of the injector: "Permission had been obtained from the Locomotive Superintendent of one of the principal Railway Companies in Great Britain to try one on a goods engine, and for me to accompany it on its trial trip, with a loaded goods train, about 70 miles out and 70 miles back. On the outward journey I was only permitted by the driver to make short intermittent trials of the Injector, he depending for his water supply upon his pumps. When we got to the end of our outward journey, and while driver and firemen were having their mid-day meal at a local public house, I went to the running shed, filled up the boiler with the injector and took out the balls from the two suction clacks and put them in my pocket. We had not gone many miles on our return journey when water was wanted in the boiler, but upon the pumps being tried, first on one side, then on the other, and naturally refusing to work without suction check clacks, I was appealed to, to put my Injector on, with the result that we completed our journey without delay or hitch of any kind, depending solely on the one No. 8 Injector. The driver consequently reported 'Pumps out of order and could not have got along without that Injector.' This was a grand testimonial, but I got into a jolly row for my temerity in removing the clack balls."

The Injector was introduced in the United States by Wm. Sellers & Co., who commenced its manufacture in 1860 at their works in Philadelphia. Of locomotive builders, Matthias Baldwin, was the first to use the new instrument, applying, in September 1860, a No. 8 Injector to an engine designed for the Clarksville and Louisville R. R. The following month the Detroit and Milwaukee R. R. put the Injector in use on one of their locomotives, and the Pennsylvania and the Philadelphia and Reading followed in the latter part of the same year.

To Jos. R. Anderson & Co., Richmond, Va., a No. 4 Injector bearing progressive number 1, was shipped in October 1860. As indicative of the wearing qualities of these early instruments Messrs. Wm. Sellers & Co. state that there was returned to them, in 1887, a No. 4 Injector, progressive No. 7, after a nearly continuous service of 27 years, and having required but few repairs; it further is interesting to note, that, owing to improvements recently introduced, American Injectors are now extensively used in France, and have been adopted as a standard type by several of the government railroads in the country of its inventor.

It need hardly be said that the Injector is the most popular boiler feeder now in use. There have been more than 500,000 manufactured in this country for the various kinds of service, and there is scarcely a locomotive in the world that is not equipped with one or two Injectors. Compact, reliable and economical, it still deserves the high encomium bestowed upon it in 1859, by M. Ch. Combes, Inspector General and Director *L' Ecole des Mines*,—" It is without doubt better than all devices hitherto used for feeding boilers, and the best that can be employed, as it is the simplest and most ingenious."

CHAPTER II.

DEVELOPMENT OF THE PRINCIPLE.

GIFFARD having established beyond doubt the power of a discharging jet of steam to lift a mass of feed water many times its own weight and force it against the initial pressure, it became necessary to prepare the constructive details of the new boiler feeder. The arrangement decided upon, could not in the light of subsequent events be considered as an entire success, as it contained inherent defects that caused frequent failures and prevented the placing of as much confidence in the new boiler feeder, as the merits of the invention deserved. Many locomotives that at first were equipped with two injectors, were afterward altered so as to have a pump upon the left hand side to be used in case the injector should refuse to work, and it was not until 1875 or 1876 that more recent improvement in construction restored the confidence that the original defects had forfeited, and the pump was driven from service upon locomotives in the United States; even yet upon some of the English Railways, a pump is used on one side of the engine, arranged somewhat in the manner of the pressure or vacuum pump for the air brakes.

The curves of the tubes and nozzles as laid down by Giffard were beyond criticism, and are still used when this type of injector is manufactured; his thorough knowledge of the laws governing the action of the jet and the accelerating velocity of the moving mass, enabled him so to construct the curves of approach and recession, that they have been used as a prototype for all subsequent forms of injectors; except for one change advanced by our increased information regarding the action of steam during expansion, and a few minor modifications for economy of manufacture,

DEVELOPMENT OF THE PRINCIPLE.

or for adapting the injector to special purposes, no change in the contour of the tubes has been made.

But that there has been development, cannot be denied; it may be considered as following three lines:

First. Constructive changes.

Second. Carrying out the ideas suggested by Giffard in his pamphlets or patent specifications.

Third. The discoveries of new properties of the jet, or the application of new principles.

Almost all important inventions follow in this natural sequence during their development, and the injector was no exception to the rule. Genuine mechanical ability is seldom combined with inventive genius, and it almost always follows that the fullest development is obtained in other hands than those of the original inventor. The first improvements were therefore in the line of correcting the defects that became apparent after the injector had been subjected to the test of actual service; changes required to facilitate repairs, or the adjustment of the positions of the tubes. In the second division lies the basis of many subsequent improvements that have since proved valuable, and Giffard has never been given sufficient credit for his wonderfully wide grasp of the possibilities or future development of the injector. Of the third there will be less to relate, as the only real advance has been with the discovery of the peculiar property of the moving jet by which the instrument was made self-regulating, and with the novel arrangement of tubes by which the re-starting feature was added.

The general appearance of the injector as now constructed is entirely different from the original form, and it would be difficult for any one not specially familiar with the subject to recognize one made in 1858; the arrangement of the adjusting handles, peculiarly shaped body, and queer little peep holes present to the modern eye a very odd appearance, while the heavy flanged pipe connections and steam cock do not contrast at all favorably with the neater form now used on American boilers.

Figure 1 shows a sectional view of the earliest form of in-

THE GIFFARD INJECTOR.

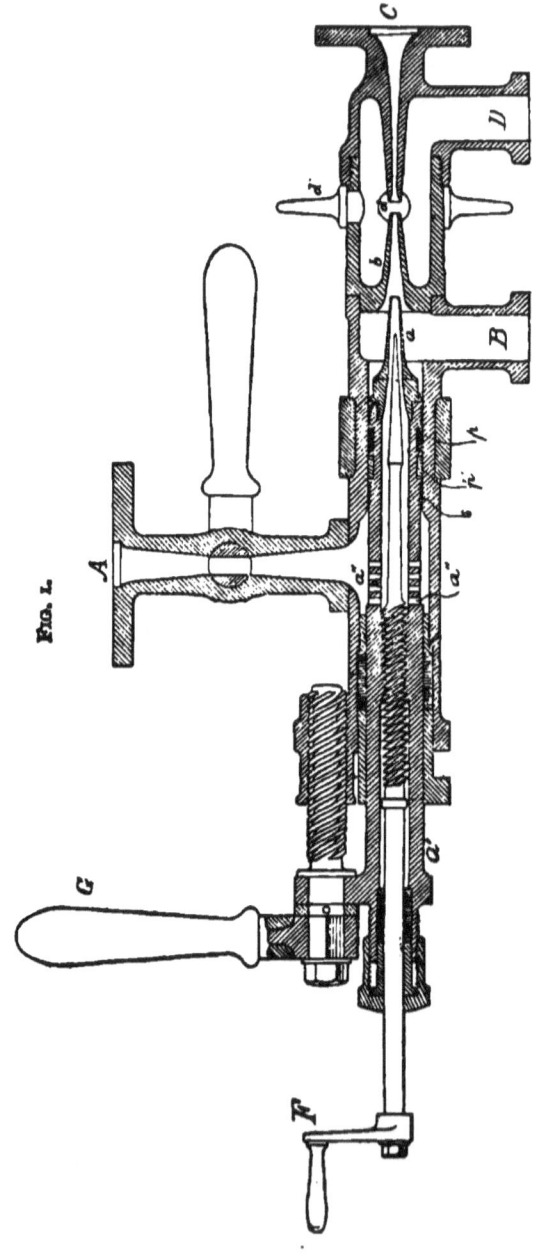

Fig. 1.

EARLIEST FORM OF THE GIFFARD INJECTOR.

jector manufactured for public sale, and was intended for use on either stationary or locomotive boilers; it was made entirely of brass, with the body composed of three pieces, screwed and bolted together, and the steam, feed, and boiler connections terminating in flanges, as is still the general practice in England and on the continent. The method of starting and operating the injector was as follows: upon opening the cock A, steam from the boiler passed freely into the steam ram a' through the small drilled holes a''; the spindle, which had been previously forced down to a tight bearing against the taper steam nozzle a, was drawn back one turn of the handle F, exposing an annular area approximately one-quarter that of the end of the tube b. The rapid discharge of the steam through the combining tube b, entrained the air in the suction branch B, and formed a partial vacuum in the feed pipe which raised the water to the injector; the spindle was then drawn fully back, and the increased discharge of steam imparted to the surrounding body of water, during its passage through the combining tube, sufficient velocity to cross the overflow space d, and enter the boiler through the delivery tube c.

Assuming the correct relation to exist between the discharge area of the steam nozzle and the annular entrance area for the water at the larger end of the combining tube, the steam will force into the boiler many times its weight of water, without tendency to spill at the overflow d; if change take place in the pressure of the steam or of the feed water, one of two conditions will be introduced, depending upon the direction in which the changes of pressure occur: either there will be an insufficient supply of water for perfect condensation during the passage of the jet through the combining tube, or too great a quantity of water will enter the injector to pass through the delivery tube c. It was to allow for this, that Giffard introduced two adjustments: one for the water area by giving an axial movement to the ram and steam nozzle by the lever G, and the other for the steam discharge area by the handle F on the spindle; if the pressure of the steam or the lift of feed water were increased, it was necessary to draw the steam ram further back to permit the admission of a larger quantity of water, while a fall of pressure would require a reverse movement. This necessity for re-adjustment was so frequent upon locomotives where the pressure was subject to wide fluctuation, that the packing upon the steam ram soon wore loose and gave constant trouble, allowing leakage to occur between the steam and feed chambers, impairing seriously the efficiency of the injector and its power of suction. This packing was formed in an ingenious manner, although in service it was found ineffective, and no device has yet been introduced that has proved thoroughly reliable for this purpose. The ram was provided with three kinds of packing: the lower part was first turned slightly larger than the bore of the body, and small triangular grooves s turned, not spirally, but in planes normal to the axis; the ram was then driven to place, forcing the tops of the sharp edges backward and reducing it to the same size as its bearing. A deeper groove p', was filled with rubber or hemp, and a second groove p, with a split metallic ring; while another set of V's was placed between the ring and the steam nozzle. Upon the body of the injector and directly opposite the overflow d, four peep holes were placed and covered by a sliding ring with handles d'; this could be rotated until the holes in the ring were opposite those in the body, and permitted the inspection of the condition of the passing jet. When new and in good condition this injector worked very well and its double adjustments gave it long wearing power; any ordinary increase in the diameter of the nozzle due to the attrition of the rapidly moving jet, could be counteracted by a change in the areas, although at a reduction of the efficiency.

As the constructive defects of the injector were obvious, the first changes introduced were improvements in the packing of the steam ram, and those manufactured in this country in 1860 were provided with a conical stuffing box filled with a number of split rings made of metal softer than the body; but unequal wear upon the ram soon caused leakage with this system also. It is singular that the merits of the fixed nozzle injector were not earlier appreciated. Millholland, of Reading, Pa., patented in 1862, an injector formed of non-adjustable tubes, depending upon the use of external valves for regulation; but this met with no success, and does not appear to have ever gained publicity. Internal adjustments seem to have been regarded as all important, for the early Giffard injectors were not provided with check valves upon the overflow, and any reduction of the water supply would cause considerable indraft of air to the boiler unless counteracted by corresponding reduction in the steam discharge; the avoidance of the check valve may have been intentional, and due to the weak form of lifting jet employed, as the power of suction was affected by the slightest increase in resistance to the free discharge of the steam. Subsequently, fixed nozzle injectors came into extensive use and are now applied to many kinds of service.

As the problem of successfully packing the steam ram seemed impossible of solution, Robinson & Gresham of Manchester, England, attempted in 1864, the plan of regulating the water area by moving the combining and delivery tubes toward or from a fixed steam nozzle; a pinion meshing into a rack cut upon the exterior of the tube was actuated by a hand wheel on the body, and a long cylindrical bearing upon the outside of the delivery tube prevented any serious leak from the boiler pipe. A similar device was introduced in 1868, by Samuel Rue of Philadelphia, who applied stuffing boxes to the ends of each tube, and regulated their position by means of a hand-lever; the fundamental problem, was, however, to make the adjustments of the water supply automatic, and eliminate as far as possible the necessity for watchful care on the part of the engineer. Giffard had already described in his patent specification two methods by which this could be obtained; one by placing a spring behind a movable steam nozzle in such a manner that an increase of pressure would cause compression of the spring, and elongate the distance between the tubes, while the resilience of the spring would regulate properly for a converse condition. This was never put into practical operation, and it is easy to see difficulties that would discourage the attempt. The other suggestion was to vary the pressure of the entering water by subdividing the steam jet and using the first or subsidiary apparatus, for feeding the second or forcing set of tubes under a pressure that would vary in proportion to the requirements. A still simpler method was that patented by Wm. Sellers of Philadelphia in August 1865, by which any change in the internal condition of the jet would actuate a set of movable tubes so that the correct proportion of the water to the steam would always be obtained.

Ever since the introduction of the Injector, the advantage of an automatic adjustment was realized, and a series of experiments had been carried on by Wm. Sellers & Co., for the purpose of attaining this end; after numerous experimental devices had been thrown aside the injector shown in section in Fig. 2, was tested and found to work satisfactorily. The basis of the invention was the discovery of the strong vacuum produced by the moving jet in a closed overflow chamber when the water supply was reduced below the maximum, and that an excess of feed water would cause the pressure in a confined overflow chamber to rise above that of the atmosphere before the jet would break. Now as the most efficient performance of an injector is when the pressure of the jet while crossing the overflow is the same as that of the atmosphere,—indicating that the density is approximately unity and that the steam entirely condensed,— it only remained to compel the condition of the overflow to influence the water entrance to the combining tube. Any variation of the steam and water supply would then first affect the absolute pressure in the overflow chamber, and this would immediately respond upon the water admission by a proper movement of the combining tube.

Fig. 2 gives a sectional view of one of the earliest experimental forms of this device with constructive details intentionally omitted. Steam, feed, and boiler connections are indicated by the letters A, B, and C, and the steam nozzle, combining tube and delivery tube, by a, b, and c; D is a closed

DEVELOPMENT OF THE PRINCIPLE.

overflow chamber, communicating with the interior of the tubes through the overflow d, situated slightly above the minimum diameter of the bore of the tube. The delivery tube c passes completely through the delivery chamber C, and is packed by stuffing boxes g and h of equal diameter, and is therefore balanced with reference to any pressure carried in the boiler, so that the tubes are free to move under the influence of any pressure within the chamber D, acting upon the differential areas of the piston b' and outside of the delivery tube c. As discharge from the overflow chamber was not permitted, the injector was started by means of a valve placed in the boiler branch C and not shown in the illustration. Two years later it was discovered that the pressure in D was strong enough to obviate the use of the balanced

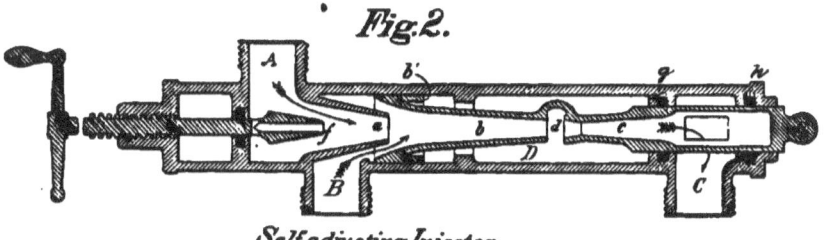

Fig.2.

Self-adjusting Injector.

delivery tube, and the simpler form shown in Fig. 3 was devised. It could be placed in any position, horizontally or vertically, as gravity had no appreciable effect upon its action. Using the same nomenclature as before, a is the steam nozzle, b the combining tube and c the delivery tube; the operation was as follows: the starting valve D, placed below the delivery tube was opened and free discharge given for the lifting steam jet issuing from a small hole drilled in the steam spindle. This gave a strong suction vastly superior to the lifting power of the original Giffard injector, and being in excess of the requirements of the ordinary height of lift, produced a lower pressure in the closed chamber d than in the feed pipe. The combining tube would therefore move down to the lower end of its stroke exposing full admission area for the water, until the rising of

GIFFARD INJECTOR.

Fig. 3.

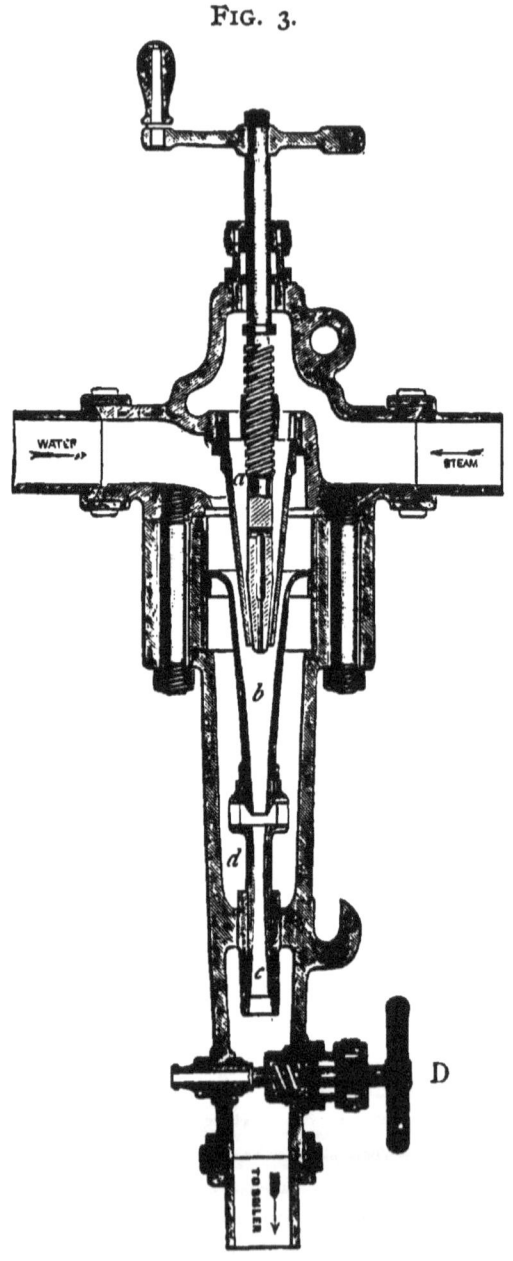

the water and condensation of the issuing steam jet caused a nearly perfect vacuum, which drew the tube to the other end of its stroke. As the spindle was drawn slowly back, the proportion of steam became too great for immediate condensation, and the contraction of the jet continued while crossing the overflow; the ensuing vacuum in the overflow chamber, acting upon the piston head of the combining tube drew the tube away from the steam nozzle, admitting an increased supply of water. It follows that regulation would occur with any change in the steam supply, whether that change be produced by a variation of the pressure of the boiler, or of the effective area of the steam nozzle. When the spindle is drawn fully back, the starting valve D is closed and the feeding of the boiler established. It follows, therefore, that a reduction in capacity of the injector may be effected by running the spindle part way down into the steam nozzle, and the water supply will be automatically reduced in proportion; at any fixed steam pressure the temperature of the water delivered to the boiler will be nearly the same for both the maximum and minimum capacity; a change in the height of lift would diminish the quantity of feed water, until a downward motion of the combining tube, due to the increased vacuum in the closed overflow chamber, would compensate by increased area, the loss of velocity due to the reduction in the difference of pressure between the interior of the combining tube and the height of lift. This method of regulation gives exceedingly good results over a wide range of pressure, and with an expenditure of steam that is proportional to the work performed.

From an inspection of Fig. 3, it will be seen that the performance of the injector depends upon the tightness of the joint between the delivery tube b, and its sleeve bearing, just as the original form relied upon the packing of the adjustable steam nozzle. Unequal wear of this part permitted leakage between the boiler and the overflow chamber, although not as rapidly nor to so fatal an extent as in the former case.

This led to the application of the discovery previously

made by Gresham, that the overflow space between the combining and delivery tubes could be lengthened without materially affecting the efficiency, and in 1873 this form of injector was still further improved by separating the combining tube from the delivery tube, screwing the latter into the body, and permitting the combining tube free movement toward and from the steam nozzle; this altered the length of the overflow space with every change in the position of the tube; with low pressure, the combining tube was drawn close to the steam nozzle, exposing the maximum length of overflow space; as the pressure rose, this distance was reduced and the length of the overflow at the higher pressure was less than that used in the older form of injector.

The principle and development of this invention has been explained in detail on account of its merit and originality. In an improved and modified form this injector is extensively used at the present time, and still preserves the peculiar principle that made it such an advance over the earlier instruments employed.

The other method of automatically regulating the water supply, by subdividing the actuating steam jet, was introduced in the United States in 1876–7, when two American patents were issued; one to Ernest Körting of Hanover, Prussia, and the other to John Hancock of Boston; foreign patents had already been granted to the former, who had been a manufacturer and experimenter for many years. The devices were practically alike; the invention consisted in combining two sets of tubes, and each set was proportioned to the function it had to perform. A small steam nozzle discharged into a large combining tube, which delivered the feed water under a slight pressure to the entrance of the smaller combining tube of the forcing set, where, meeting the steam issuing from a larger nozzle, the water received an additional impulse sufficient to force it into the boiler; the increase in feeding capacity of the first set of tubes with a rise in the steam pressure, obviated the necessity for any regulation of the water supply to allow for fluctuation in the pressure carried in the boiler.

DEVELOPMENT OF THE PRINCIPLE.

A section of this injector is shown in Fig. 4 where a and b are the tubes of the second, or forcing set, a' and b', of the lifting set; an overflow cock is placed beyond each delivery tube for the purpose of starting, as no overflow aperture separates the convergent combining tube from the divergent delivery tube. The proportions in which the sets differ are apparent; the first set is primarily, an ejector, as the diameter of the steam nozzle is less than that of its delivery tube. In the other set the conditions are reversed, and an injector is formed which is capable of forcing the feed water against a pressure considerably above that of the initial jet. A separate valve is provided for each steam nozzle, so that

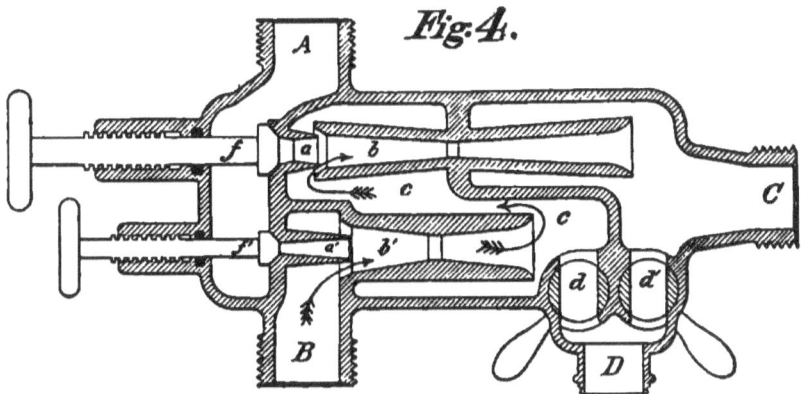

by admitting steam first to a', the feed water is raised and flows from the waste cock d'; this is now closed, compelling the water to pass through the smaller combining tube b', the steam nozzle a is now opened and the jet established through the forcing combining tube, aided by the velocity of the entering water due to the pressure produced by the delivery tube of the lifting set; the closing of the final waste cock d compels the water to enter the boiler.

The self-adjusting principle of the Double Jet Injector depends upon the variation of feed pressure when the water entrance to the combining tube is constant. Numerous changes have since been made in the special devices by which the necessary sequence of opening and closing the various steam and waste valves could be obtained. In its modern form, this injector is very convenient to operate, and the principle of its construction is incorporated in many of the best known injectors of the present day.

In tracing the development of the self-adjusting principle, several other improvements of the more elementary forms have necessarily been omitted; these will now be taken up as nearly as possible in the order of their introduction.

In this country the self-adjusting injector was so early perfected, and it so largely superseded the original form, that few other improvements were made until the approach of the expiration of Giffard's patents. In England and France, the form of the injector as manufactured by Sharp, Stewart & Co., and H. Flaud & Cie., was modified in various ways; Stewart, Robinson and Gresham, brought out several meritorious inventions. Among the most important of these were the adjustable combining tube, which has already been described, and the separation of the lifting apparatus from the injector proper; in the latter case an ejector, whose only function was to raise the feed water, was connected with the overflow aperture of the combining tube, and by its aid the injector could be started with great facility. In 1863 the same firm patented an arrangement by which the injector was placed below the level of the water in the tank of a locomotive; the waste pipe from the overflow extended up into the cab above the water level, and any waste from the injector could be readily seen and corrected; further, the drip pipe was connected with the suction, and loss of feed water from any cause prevented.

In France, the few improvements that were added, were chiefly in the line of English models. Turck followed Gresham in the use of a stationary steam nozzle, and in addition applied an enveloping nozzle, enclosing an air space, in order to prevent transmission of heat from the steam to the surrounding feed water. Cuau brought out an injector in which all the tubes were fixed, and so proportioned as to permit the use of water of very high temperature; no lifting spindle was provided, but the feed water was raised by an auxiliary ejector. Bouvret, Polonceau, and Delpeche, followed with comparatively unimportant modifications, although all have lent their name to styles used upon railroads in France.

As no patent was allowed Giffard in Austria and Germany, his injector was manufactured by unlicensed parties soon after its introduction in France, and much original investigation was started. Schau discovered the advantage of the divergent steam nozzle in 1869, by which the expansion of the steam is utilized to better advantage than with the convergent shape. Haswell, Körting, Pradel, and Krauss, introduced special changes, many of which are still in service. Körting, whose invention of the double jet injector has already had special consideration, was granted a patent in 1872 for drawing exhaust steam from the cylinder into the injector, by which the temperature of the water delivered to the boiler was considerably raised; an opening was made in the combining tube at a point where a strong suction was produced by the condensation of the motive jet: the inventor states in his specifications that the injector could be used with live steam alone, and that the supplementary overflow enabled it to start very easily. Friedman in 1879, patented an improved fixed nozzle injector, and in the form in which he placed it upon the market, used a supplementary opening in the combining tube for the entrance of an additional supply of feed water; this system possessed a number of advantages, and after undergoing subsequent modification is now extensively used both in Germany and the United States.

In England in 1877, Hamer, Metcalf & Davies, obtained a patent for an injector in which there was a very large steam nozzle, and a combining tube with a hinged section extending nearly its full length; these modifications were made for the purpose of using exhaust steam for the actuating jet, and the injector was enabled to feed boilers carrying as high as 80 lbs. pressure; the tubes were well designed and showed much ingenuity in their construction.

Returning to American improvements we find Garner C.

Williams of Ellenville, N. Y., made in 1880 application for a patent for the first self-starting injector;* it was very ingeniously constructed, but seems never to have appeared upon the market. John Loftus, of Albany, N. Y., followed in 1881 with a simpler device, shown in Fig. 5, which differed from all previous forms by the insertion of a cylindrical suction or draft tube b', between the steam nozzle and the combining tube. This obviated the necessity for a special priming device, as the opening of an ordinary globe valve in the steam pipe would produce a vacuum in the water branch sufficient for all ordinary lifts. Check valves $h\,h'$ were placed on the separate overflow chambers D, D', so that the condition of the jet while crossing the starting

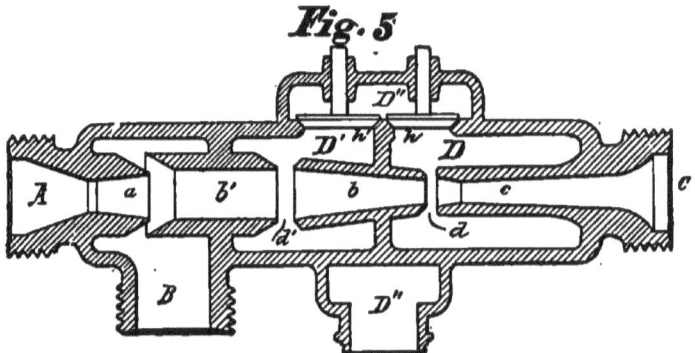

overflow d, was independent of, and uninfluenced by, its condition at the upper or supplemental overflow d'. Bancroft, Gresham, Penberthy, Desmond, Derby and others, altered and improved the arrangement of the valves and tubes of the re-starting injector and increased the efficiency of its performance.

In 1885, J. Sellers Bancroft of the firm of Wm. Sellers & Co., of Philadelphia, patented a modification of the double jet principle, in which the re-starting feature was introduced; starting and supplementary overflows were so placed in the combining tubes of the forcing and the lifting sets, that free discharge from the steam nozzles was obtained, producing a strong vacuum in the suction pipe. This form

* Subsequent research shows that a re-starting injector was patented in France by Vabé, Nov. 12, 1878.

DEVELOPMENT OF THE PRINCIPLE. 23

of injector was not placed upon the market, as further experiment by the same firm developed the arrangement shown in Fig. 6, invented by the author in 1887. This injector consists of a double set of tubes arranged in axial line, giving continuous acceleration to the jet from the moment the water enters the draft tube. A, B, C, are the steam, feed, and boiler connections; a', an annular lifting steam nozzle discharging through the annular draft tube b', and a the forcing steam nozzle from which the jet receives the final impulse sufficient to enter the boiler. The overflow openings d, d', d'', are all contained in a chamber D, communicating with the large waste pipe D' only by the opening of the waste valve h: these overflow apertures are of such proportion and so distributed that free discharge is obtained for

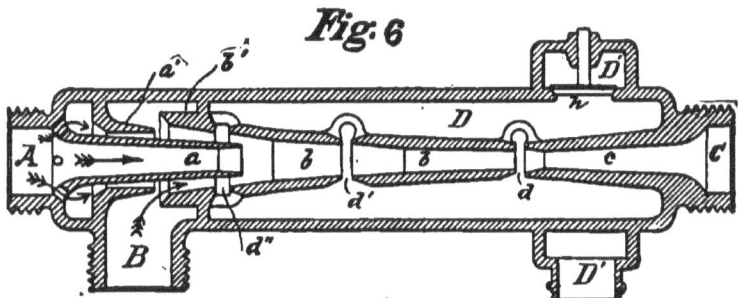

the lifting steam nozzle a'. The resultant vacuum in the feed pipe always tend to raise the water to the injector after a temporary interruption of the water supply and restore the continuity of the jet, while the arrangement of the tubes and steam nozzles permit wide variation of steam pressures without waste at the overflow.

In 1890 another form of locomotive injector was invented by Holden and Brooks, of Salford, England, who had already secured several patents for fixed nozzle exhaust steam injectors; this device combined the use of exhaust and live steam, with the tubes arranged as in the double jet, and so contrived for automatic action, that opening valves in the feed and exhaust pipe would start the injector even against the high pressures carried in locomotive boilers.

There have been numerous other changes made upon the invention of Giffard, but those that have been described, represent steps in the development of the injector; they have been selected from over five hundred patents granted by the United States, and as many more by foreign countries. Many changes may seem to be merely matters of detail,—yet it has only been by continued experiment with modification of minor parts, or slight changes in the proportions or arrangement of the tubes, that progress has been made. The original invention was conceived upon theoretical principles, the correctness of which has been so signally demonstrated, but since that time the greater part of the work of development has been attained only by close attention combined with critical observation in experimental research.

CHAPTER III.

DEFINITION OF TERMS—DESCRIPTION OF THE IMPORTANT
PARTS OF THE INJECTOR—THEIR FUNCTIONS.

BRIEF descriptions have already been given of the different types of injectors in connection with the history of the development, but as the technical names of the various parts will necessarily enter into the accounts of experiments that follow, the definition will now be given as concisely as possible.

It is unfortunate that there has been so little accord in the direction of injector nomenclature; patent specifications are sometimes obscure and misleading, and the terms used often confusing; even the names ejector and injector are often appropriated to entirely wrong uses. The two apparatus differ in principle as well as in essential detail: the former can operate with any gas or any liquid in conjunction with any other gas or liquid, while the action of the injector depends upon the expansion and condensation of a gaseous fluid acting within clearly defined limits. Technically,

An Injector is an apparatus in which a gaseous jet impinges and is condensed by a fluid mass whose final kinetic energy exceeds that of a jet of similar form and density discharging under the initial pressure of the motive jet.

In addition to the use of condensible gases, the existence is required of certain well established relations between the areas of the discharging and the receiving nozzles, which will serve for a clear distinction between the two types of apparatus:

An Injector is a jet apparatus in which the cross-section of the discharge nozzle of the actuating jet is greater than that of the receiving or delivery tube.

An Ejector, reverses these conditions, and the gaseous or liquid discharge will freely pass through the delivery tube.

Drawings or cuts of injectors are usually made with the delivery tube discharging toward the bottom or right hand side of the page, and the terms "upper" and "lower" will be thus used. The index letters refer to the various figures, where the same letter always is used on the same tube.

The Delivery Tube (c) is that tube in which the maximum velocity of the combined mixture of water and steam is attained, and subsequently reduced, by means of the expanding curves and increasing cross-section, to the velocity and pressure in the boiler pipe.

It is usual to indicate the nominal size of the injector by the mininum diameter of this tube, as the amount of water delivered is chiefly dependent upon this dimension.

The Combining Tube, (b) extends from the upper end of the delivery tube (c), to the lower end of the steam nozzle (a), and receives its name from the combining of the steam with the water that occurs within its walls.

The Draft, or Suction Tube, (b') facilitates the starting of the injector, and lies between the upper overflow (d') and the lower end of the steam nozzle (a); its lower diameter is usually larger than the minimum diameter of the steam nozzle.

The Steam Nozzle (a), guides and directs the motion of the actuating jet; its effective area of discharge is often varied by the use of a *Steam Spindle* (f), in order to vary the amount of steam used, or for the purpose of lifting the feed water.

These tubes can be used in either single or double combination as in the single or double jet injectors.

Overflow, (d) Primary or Lower, a narrow annular vent space or drilled aperture, placed above the minimum diameter of the delivery tube, permitting free outlet for the water and steam during the operation of starting.

Upper, or Supplemental (d'), is placed nearest to the steam nozzle, and often made of sufficient area to permit free exit for the full discharge of the steam nozzle.

Additional overflows are frequently used, see Fig. 6, page 23 ($d''\ d'''$).

The following terms are used as descriptive of different classes of injectors:

Single Jet Injector, one in which a single set of combining and delivery tubes is used.

Double Jet.—An injector containing two sets of steam jet apparatus, of which the first or lifting set receives the feed water from the source of supply and delivers it to the second, or forcing set, from which it receives sufficient impulse to enter the boiler.

Automatic or Re-Starting.—An injector that is able to re-establish automatically the continuity of the jet, after a temporary interruption in the steam or water supply.

Self-adjusting.—Applied to an injector in which the supply of water is automatically adjusted to suit the steam supply without waste at the overflow.

Open overflow injector, has one or more apertures in the combining tube, opening into one or more overflow chambers, that may be closed against the admission of air by the use of light check valves opening outward.

Closed overflow injector, can only start by means of an opening or vent placed beyond the delivery tube, which must be closed in order to divert the jet into the boiler.

The following terms relate to the performance:

Maximum capacity.—The greatest volume or weight of feed water passing through the delivery tube at any given steam pressure and condition of feed. It is usually measured in *cubic* feet or *pounds* per hour; the use of the gallon is not to be advised, unless its value is clearly stated.

Minimum capacity.—The least volume or weight of water as above, that can be continuously delivered against boiler pressure without waste from the overflow. It is often expressed as a percentage of the maximum capacity.

Range of capacities.—The difference between the maximum and minimum capacities expressed in terms of the maximum capacity; for example, if the max. is 300 cu. ft., and the min. is 200 cu. ft., the difference, 100, is the range, which is 33 per cent.

Overflowing temperature.—Highest admissible temperature of feed water with which the injector can operate without wasting, when running against boiler pressure.

Overflowing pressure.—Under given conditions, the highest counter pressure against which the injector can run without wasting.

Breaking temperature, Breaking pressure.—Highest admissible feed temperature, and counter pressure, when the waste valve is closed. *Note.*—With closed overflow injectors, the overflowing and breaking temperatures and pressures are in most cases the same.

Efficiency.—This may be based upon the ratio which the total heat in the feed water and in the steam, bears to the heat in the delivered water plus the heat equivalent to the work of forcing the water into the boiler. This may be called the "Thermodynamic Efficiency." Or, the ratio of the work performed by the steam in forcing the water into the boiler, to the total energy given out by the steam during its expansion from the initial boiler pressure to the pressure in the combining tube. This is the "Mechanical Efficiency." Or, it may be expressed in terms of the weight of water delivered per unit weight of steam; this is a very simple and convenient methed of comparison, and is thoroughly practical.

Note :—The efficiency is often measured by the volume or weight of water delivered per unit area of cross-section of the steam nozzle or the delivery tube.

DEFINITION OF TERMS.

The description of the tubes of the injector will be taken up in the following order, Delivery tube, Combining tube, and Steam Nozzle, the theory of their action reviewed, and practical questions considered that have important influence upon their design.

CHAPTER IV.

THE DELIVERY TUBE.

DESCRIPTION :—EFFICIENCY OF VARIOUS TYPES—EFFECT OF DIFFERENT SHAPES AND PROPORTIONS.

THE function of this tube is to change the kinetic energy of the jet to potential, with the least possible loss : or, to transfer the energy due to the velocity of the water and condensed steam, into pressure in the boiler pipe.

Of all its dimensions, that of the minimum diameter is the most important, for it is from this base that the dimensions of all the other parts are calculated. The use of this diameter to denote the size of the instrument was first suggested by Giffard, and this seems to be the most rational method that can be used. The quantity of water delivered by an injector is directly dependent upon the minimum area of this tube, and under the same conditions, varies with the square of the diameter; theoretically, the capacity should be equal to that quantity of water which would be discharged from a similar orifice under a pressure or head equal to the overflowing pressure of the jet, which is always in excess of the pressure carried in the boiler. If there were no losses, this would give an exact method of determining the weight of water delivered, and as it is, the losses referred to, can be closely approximated; by comparing the calculated discharge with the actual test of the injector, a percentage of efficiency on this basis can be readily determined.

This basis of efficiency has, in the cases of all single jet injectors, a decided value; the smaller the delivery tube, the smaller can be the other orifices of the injector, reducing the weight of steam used, and rendering its operation more economical. As an example, take a case from actual practice, from which the efficiency will be calculated.

A No. 8 Injector, having a delivery tube 8 millimetres in diameter, commences to waste against 160 pounds (corrected) back pressure; the delivery temperature is 148 deg. Fahr.

The height of a column of water at this temperature that corresponds to the pressure of one pound per square inch, is 2.354 ft., (see Table II., page 43) and the head of water equal to 160 lbs is 2.354 × 160 = 376.64 ft. This gives a velocity,

$$v' = \sqrt{2gh} = 8.024 \sqrt{h} = 8.024 \sqrt{376.64} = 155.721 \text{ ft. per sec.}$$

The area of the delivery tube in square feet is 0.000541068, and the volume that would be discharged *from* this tube under this head

$$155.72 \times 0.000541068 \times 3600 = 303.3197 \text{ cu. ft. per hour,}$$

which must be equal to the theorectical quantity of water entering against that pressure. But an actual test showed that 251.5 cubic feet had been taken from the feed tank, and that the weight of dry steam used was 1304.5 pounds; therefore the total volume passing through the delivery tube per hour was

$$251.5 + \frac{1304.5}{61.244} = 272.8 \text{ cu. ft.}$$

The density of the jet was

$$\Delta = \left(\frac{272.8}{303.3197}\right)^2 = 0.8087$$

and the actual velocity of the jet

$$V = \frac{155.721}{\sqrt{0.8087}} = 173.15 \text{ ft. per sec.}$$

The discharge per hour per square millimetre of cross-section

$$\frac{272.8}{50.2656} = 5.427 \text{ cubic feet.}$$

This gives a very simple means of comparing different forms of injectors, and the following table is based upon actual tests of single jet locomotive injectors at a steam pressure of 120 pounds to the square inch; no Double Jet Injectors are included, as the conditions under which the jet passes through the delivery tube of the two types are not exactly comparable:

TABLE I.

NAME.	Size.	Diameter Delivery Tube in Millimetres.	Cu. Ft. per hour per sq. m. m.
Garfield	No. 7	7.26	4.8876
Mack	No. 7	8.13	3.8006
Monitor	No. 8	8.28	5.1258
Metropolitan	No. 9	10.75	3.8555
Original Giffard	No. 8	8.00	4.4112
Sellers' 1876	No. 7	7.00	5.6745

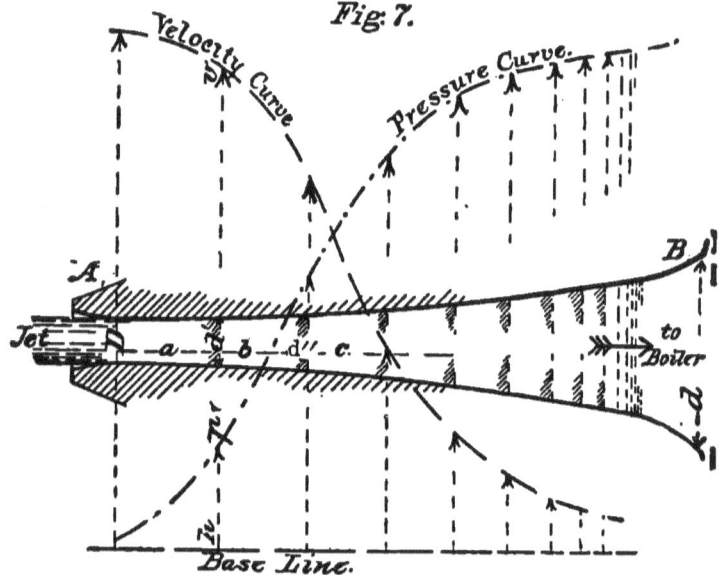

Fig. 7.

The principal function of the delivery tube has already been stated, but it remains to show the means by which the required effect is produced. A section of a tube is shown at AB in Fig. 7, where, for the sake of clearness, imagine the entering jet divided into a series of short cylinders, of the

THE DELIVERY TUBE.

same density as water; also, that the injector is working against a pressure that compels the tube to be completely filled with water; now, if the tube were cut off at the section D, the jet would impinge violently against the wall of water in front, and lose its energy in forming eddies and disturbances, (see Fig. 8), in proportion as its velocity is greater than that of the water ahead. But as the jet passes into the tube, the cross section becomes wider, and the cylindrical form changes to conical. If the volume contained between any two sections as Dd', or $d'd''$, etc., be the same, the distances a, b, c, gradually shorten as lateral motion toward the walls is induced; as these distances shorten an increasing pressure is exerted upon the surrounding walls and upon the particles of water directly in front. In this way the momentum of the jet, which is at the maximum at D,—as indicated by the highest point of the velocity curve,—is gradually reduced to that of the feed in the boiler pipe. By reference to the figure the pressure and velocity of the jet at any point can be determined by the height of the corresponding curve above the base line. If the density of the jet be uniform, its velocity at any point of a given tube can be obtained by dividing the actual volume passing through, expressed in cubic feet, by the area of the tube in square feet, and the side pressure upon the walls may be calculated by the following formula:

$$p_1 = \frac{y}{144} \left[\frac{144\,P}{y} - \frac{1}{2g}(v_1^2 - v^2) \right] \quad \ldots \ldots \ldots (1)$$

Where p_1 = pressure required.
P = pressure in boiler pipe at end of the delivery tube.
v = velocity of water in boiler pipe.
v_1 = velocity at point where pressure = p_1.
γ = weight of one cubic foot of water at the temperature of delivery.

It is of course advantageous to shape the tube so that all the energy of the jet can be utilized, but this can only be done by applying the laws governing the motion of fluids; the effects produced by improperly shaped tubes may be seen by the following tests, where the same conditions obtained throughout, both as to steam pressure and to the volume of water passing through the injector. The first form experimented with is shown in Fig. 8, which corresponds to the mouth or entrance of a delivery tube, as if the part to the right of the section D, in Fig. 7 had been taken entirely away; the result of this was that the injector would

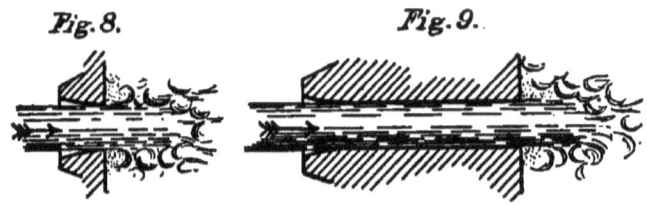

Fig. 8.   Fig. 9.

only force against a pressure of 35 pounds without overflowing, although the pressure carried in the boiler was 65 pounds to the square inch, showing the enormous proportion of energy dissipated.

The next tube was made cylindrical, as is indicated in Fig. 9. Under the same conditions this tube showed no improvement, as only 25 pounds was reached; it was then

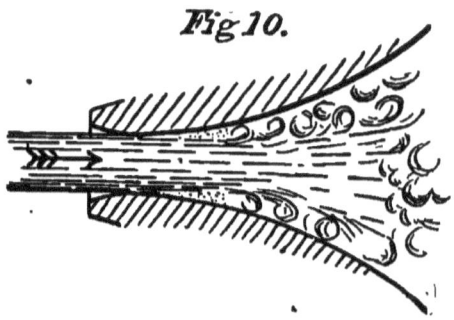

Fig. 10.

reamed in the form of a divergent curve approximating in section a parabola as shown in Fig. 10, and the effect of the change was at once apparent as the permissible back pressure rose to 62 pounds; although the general shape was improved, yet the tube was obviously too short and the curvature too great for the high velocity at which the jet was moving, and the tube in Fig. 11 was substituted. Its length was 7.6 times the diameter, and developed a pressure of 88 pounds. Further change in the proportion raised the pressure against which the jet would work to 93 pounds, without any modification of any of the other parts of the injector, or increase in the pressure of the steam.

It has long been known that the divergent tube possessed the peculiar property of increasing the quantity of water that would flow through an orifice in a given time, and this phenomenon has led to careful experimental work by Bernouilli, Francis, Brownlee, and others. Francis in his well-known and oft-quoted Lowell Hydraulic Experiments, found that with an orifice having well rounded curves of approach, the weight of water discharged under a constant head of 1.36 feet, could be increased 2.44 times by the addition

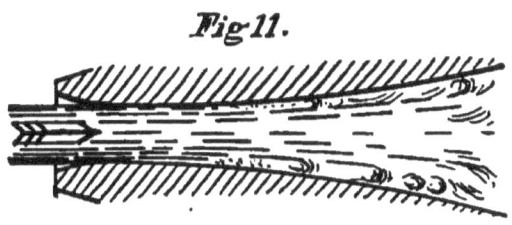

Fig 11.

of a divergent funnel having an angle of 5° 1'. Brownlee, with 6 feet head of water increased the discharge 2.42 times by the use of a tube whose included angle was 7° 5'. But the heads of water and the velocities of the jets in both these cases were exceedingly small compared with those in use in the injector, and the changing conditions in the internal action of the jet would invalidate any positive prediction regarding the effect of a special shape of divergent tube; yet the theory under which such tubes may be designed is interesting, and may be applied in a limited degree to the case under consideration, under the assumption that the jet has the same density as that of water; this is however only approached in those types of injectors which are most carefully designed and constructed.

THE GIFFARD INJECTOR.

In order to obtain the most efficient form of divergent tube, it has been suggested by Nagle that it be so constructed that the retardation of the motion of the jet be made uniform, and that by this means the particles of water could be kept in equilibrium and internal eddies avoided. This would require each succeeding section of the delivery tube to be increased in such proportion that the difference between the squares of the velocities of any two equidistant sections would be constant; and the particles of water always in contact with its walls. It should be remembered that the velocity of the entering jét must be always equal to that of a jet of similar density discharging *from* the tube under a head equal to the difference between the maximum pressure against which the jet is capable of working, and the internal pressure of the jet at the time of entrance; it follows therefore that the absolute pressure of the jet at this point would influence the amount of water passing through the tube, if the construction of the other parts of the apparatus were such as to permit an additional quantity of water to be drawn from the supply.

The reason for the lowering of the pressure at the point of minimum diameter may thus be explained: a jet of water discharging freely into air, will retain its full velocity and the same cross-secton for some distance beyond the mouth of the tube; but by reason of the gradually expanding curves of an enveloping tube, the jet adheres to the surrouning walls and its section is increased; but the energy tends to remain the same as before, and therefore an effort is exerted to draw the preceding particles forward, increasing their velocity, and consequently the weight of water discharged. If the tube is open to the atmosphere, a lack of equilibrium results as air is drawn in at the lower end, but if immersed, the effect reacts upon the particles in the rear, and, if the tube is correctly proportioned, a perfect vacuum will be formed at the throat of the tube.

The formulas for uniform retardation are similar to those for uniform acceleration, and may thus be expressed:

Let V be the velocity of the jet and D the diameter of the

THE DELIVERY TUBE.

tube at the throat; let v and d have the same values at the lower end of the tube; then

$$V : v :: d^2 : D^2 \text{ and } v = \frac{V D^2}{d^2} \quad \ldots \ldots \ldots \ldots (2)$$

If S is the length to the tube, the negative acceleration, or retardation is

$$p = \frac{V^2 - v^2}{2 S} \quad \ldots \ldots \ldots \ldots (3)$$

or

$$p = \frac{V^2 - V^2 \frac{D^4}{d^4}}{2 S} = \frac{V^2}{2 S}\left(1 - \frac{D^4}{d^4}\right) \quad \ldots \ldots \ldots (4)$$

If the lower end of the tube be very large the last term of (3) becomes so small that it may be neglected without sensible error, and may be reduced to

$$p = \frac{V^2}{2 S} \quad \ldots \ldots \ldots \ldots \ldots (5)$$

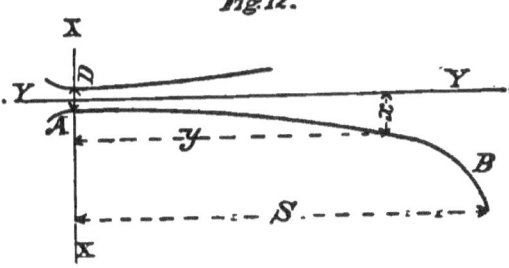

Fig. 12.

The equation of the curve of the half section of a tube that would fulfill these conditions may thus be found; in Fig. 12 let $A B$ be the half section of a delivery tube whose centre line is $Y Y$; vertical ordinates denoted by x, and abcissas by y parallel to $Y Y$; substituting $2 x$ for d, and y for S, in (4) and altering its form,

$$x = \frac{D}{2} \sqrt[4]{\frac{V^2}{V^2 - 2 py}} \quad \ldots \ldots \ldots \ldots (6)$$

By way of example, take a delivery tube 8 millimetres in diameter and the length, S taken as 20. (This means 20 spaces of equal length and each space may be any desired distance although it is advantageous to make the total length as great as possible.) The entrance velocity of the jet may be assumed as 156 feet per second, and the final velocity, v, as 0. Substituting in (4) gives the retardation, and (5) reduces to

$$x = 4\sqrt[4]{\frac{1}{1 - 0.05y}}$$

from which the diameter can be determined for any part of the tube; for the following values of y, d has been calculated, and it will be noticed that the tube widens very slowly at first, but as the lower end is approached the curve becomes very steep and finally tangent at B.

For $y =$	$d =$
0	8.000
5	8.590
10	9.512
15	10.912
20	∞

Looking at these formulas from a practical point of view, it will be seen that they depend upon the value of V, the entrance velocity, which in the case of the injector must vary with every steam pressure and change in the density of the jet, although it is at the maximum capacity, where the density most nearly approaches unity, that the full efficiency is needed. Giffard realized the advantage of using differently shaped tubes for high and low pressure, and advised the use of circular arcs whose radii in the former case were 300 times the diameter of the tube, and in the latter, 200 times. This altered the proportion of the tube and the mean angle of divergence of the tube for each size instrument, but this seems for many reasons to be theoretically correct, and an improvement upon the straight taper adopted by most manufacturers.

The change in the velocity of the jet and the gradually increasing pressure upon the walls of the tube was clearly shown for the theoretical case assumed in Fig. 7; as an illustration of the actual changes of pressure occuring within the tube under different conditions, the diagram shown in Fig. 13, is presented through the courtesy of Wm. Sellers & Co., of Philadelphia, who have, ever since their introduction of the injector into the United States, made experimental work an important part of their system. These tests were made with a Self-Acting Injector of 1887, at 65, 100, and 120 pounds pressures. The delivery tube was pierced at four points, A, B, C, D, by small drilled holes one-sixteenth inch in diameter, upon which gauges were placed to indicate the internal pressure of the jet. Under a head of 30 feet, water was allowed to discharge through the tube and the vacuum at the throat of the tube, as indicated by a U mercury gauge, was found to be $27\frac{1}{2}$ inches; as measured by actual test under a head of $18\frac{1}{2}$ feet, the amount of water passing through was increased a little over 42 per cent. above that discharged by a similar orifice without the divergent tube. The pipes were then changed so that the feed water could be lifted from a tank placed below the injector, and the steam turned on; the pressures observed were plotted on the vertical lines above the corresponding point of the tube and connected by the curved lines and marked with the initial steam pressure. The rise in pressure when the injector is running at its maximum capacity is indicated by the full line, the minimum by the dotted lines, and a capacity approximately half-way between, by the alternate dot and dash.

It will be noticed that the curves for the maximum capacity rise easily from the low pressure at the throat of the tube to the final boiler pressure at the end, distributing the wear very evenly. At the mean capacity, the point where the greatest abrasion will occur is evidently at the section marked C, where the pressure line rises suddenly from $10''$ vacuum to 95 pounds; at other steam pressures the same peculiarity was noticed, but the lines were not added to the diagram as they would have sacrificed its clearness. At the minimum capacity, the wear is chiefly between A and B, as may be seen by reference to the dotted curves. These facts constitute a strong argument in favor of attaining the per-

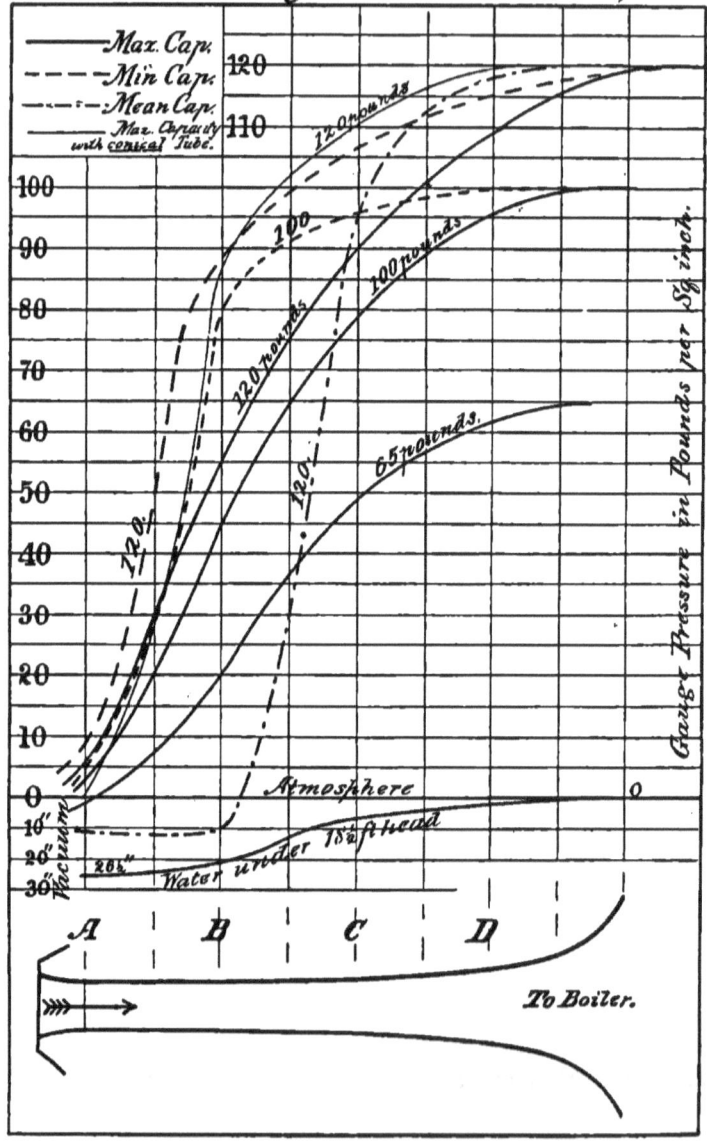

Fig. 13.
Pressure in Delivery Tube at 120, 100 and 65 lbs. steam.

THE DELIVERY TUBE. 41

fect condensation of the steam within the combining tube and before the entrance to the delivery tube is reached, and one of the reasons for basing the efficiency of an injector upon the amount of water delivered per unit of area of delivery tube.

The disadvantages of the conical tube may be seen from an examination of the light full line, marked 120 pounds; the abrupt rise in pressure against the walls of the tube, even on the maximum capacity and when all conditions are the same as in the other experiments, shows the inferiority of this form of tube.

The various causes of loss of energy of the jet can be closely approximated in the most important cases that occur in practice. That due to the final velocity in the delivery tube, and impact of the water against the slowly moving water in the delivery pipe, can be found from the expression,

$$p = \frac{\gamma}{144}\left(\frac{d_1^2}{d^2} - 1\right)^2 \frac{v_1^2}{2g} \quad \ldots \ldots \ldots \ldots (7)$$

Where p = the loss in back pressure in pounds.
 v_1 = the velocity in the boiler pipe.
 d_1 = the diameter of the boiler pipe.
 d = the diameter of the lower end of the delivery tube.
 γ = the weight of a cubic foot of water at the temperature of the delivery.

The loss in head due to the friction of the jet upon the walls of the tube, can be calculated from the formula for conical pipes; or, if curved, by using the nearest angle or angles to correspond. This loss amounts to very little, compared with the pressures carried, and this equation is much simpler than that for tubes formed from circular or parabolic arcs. In Fig. 14 let

 p = loss in back pressure due to friction.
 V = velocity of jet at the throat of tube.
 D = diameter of the throat.
 d = diameter of lower end of tube.
 δ = included angle of tube.
 β = co-efficient of friction, depending upon condition of tube.
 γ = weight of one cubic foot of water, as before.

$$p = \frac{1}{8}\beta \, \csc\frac{\delta}{2}\left[1 - \left(\frac{D}{d}\right)^4\right]\frac{V^2}{2g} \times \frac{\gamma}{144} \quad \ldots \ldots (8)$$

To apply these two formulas to an actual case, take a delivery tube 0.3" at the throat, 3.6" long, and a taper of 1 in 12 in diameter; the included angle is 4° 46". Take V as 156 feet per second, and β, as 0.0147; d is found to be 0.6"; substituting,

$$p = \frac{1}{8} \times 0.147 \times 12.078 \times \left[1 - \left(\frac{0.3}{0.6} \right)^4 \right] \frac{\overline{156}^2}{64.4} \times \frac{1}{2.354}$$

whence,

$$p = 3.5 \text{ pounds} -$$

and from formula (7),

$$p = \frac{1}{2.354} \left(\frac{4}{.36} - 1 \right)^2 \frac{\overline{35}^2}{64.4} = 9.8 \text{ lbs.}$$

Fig. 14.

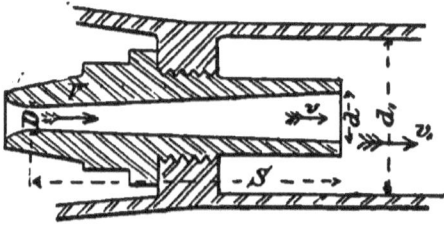

This loss could be entirely obviated by widening the end of the tube in easy curves to the diameter of the boiler pipe.

The co-efficient, β, will depend upon the condition of the walls of the tube, whether smoothly reamed and offering but little resistance to the motion of the jet, or whether abraded by constant use or cut in circular grooves by poor workmanship. When it is remembered that the velocity of the jet at the smallest diameter of the tube corresponds to that attained by a jet of the same density issuing under a pressure even higher than that in the boiler, the disadvantageous effect produced by the roughened surfaces of the guiding tube can be easily realized.

This leads naturally to the subject of the wear of the tube. From what has been said regarding the distribution of internal pressures, the points where the greatest wear will occur under usual conditions, can be easily seen. The presence of grit or dirt in the feed water will cut the mouth of the tube, and enlarge the minimum diameter, acting, by reason of the high velocity of the jet, like a continuous grinding cylinder; but it is further down in the tube that the abrading effect is first noticed, near the place marked C in Fig. 13, where an annular groove is often worn around the tube, that seriously impedes the motion of the water; its location is always at that point where the pressure suddenly rises; as the cross-section gradually widens, the velocity of the water decreases, and the wear is less noticeable. Regarding the question of repair, it may be said that this tube will generally be the first that requires replacing, but it may often be saved by carefully reaming or smoothing out the roughened places; if this be done, the minimum diameter may be somewhat increased before the injector will cease to work at the higher steam pressures. Its power at low steam will be affected first by the wearing of the tube at this point, but the exercise of a little judicious management will often prolong the life of the injector, and save needless substitution of new parts.

The following table of the weight of a cubic foot of water at different temperatures, and the head of water in feet corresponding to a pressure of one pound per square inch, will materially assist the calculation of the performance of the injector under different conditions. It has been compiled from a "Table of Comparative Volumes," prepared by Mr. A. F. Nagle, and published in *Proc. Am. Soc. Mech. Eng.;* the heads of water are based upon his figures.

TABLE II.

Temp. Water, deg. Fahr.	Weight of 1 cubic foot.	Head in feet = to 1 lb. per square inch.	Temp. Water, deg. Fahr.	Weight of 1 cubic foot	Head in feet = to 1 lb. per square inch.
39.1	62.4250	2.3067	130	61.5320	2.3402
40	62.42398	2.3068	140	61.3432	2.3474
50	62.40735	2.3074	150	61.1413	2.3552
60	62.36975	2.3088	160	60.9266	2.3635
70	62.31015	2.3110	170	60.6988	2.3723
80	62.2283	2.3140	180	60.4608	2.3818
90	62.1253	2.3179	190	60.2128	2.3915
100	62.0033	2.3224	200	59.9569	2.4017
110	61.8626	2.3277	210	59.6935	2.4123
120	61.7053	2.3336	212	59.6400	2.4144

CHAPTER V.

THE COMBINING TUBE.

IN describing the action of the jet within the delivery tube, certain theories were given which seemed to correspond closely to the most important conditions occurring in actual practice; unfortunately the action of the steam in the combining tube is not so definitely understood, and the most that can be done is to describe the phenomena as observed, and deduce a few general conclusions.

It is probable that more experimental work has been required to perfect this tube than any other part of the injector, and each investigator has adopted a special form to suit pre-determined conditions, that seem to him to be most effective. The best shape can be decided by experiment only, and therefore the special conditions under which the tests are made govern the result; it follows, therefore, that there may be as many different forms as there are manufacturers, and to a great extent this is true, each type of injector operating more or less successfully under a certain range of conditions.

The first requisite condition that the tube must fill, is that the water must be sustained during the impact of the steam, and the second, that the mixture of the water and the steam be made as intimate as possible in order that complete condensation may take place during the passage of the jet through this tube; this can only be done by using correct proportions at the upper end, and then giving to the lower or convergent part the same shape that the jet would assume during the process of condensation.

It is this tube, in great measure, that governs the mechanical efficiency of the injector. Of course each tube has its own function, yet all are inter-dependent; but the process of condensation, that differentiates the injector from other similar apparatus, occurs within the walls of this tube. In the first place, assuming a constant head of feed water, the cross-section of the upper end, by regulating the quantity of water that may enter, determines the proportion of water to steam in the resultant mixture, and the temperature of the delivery. The vacuum in the tube is dependent upon this temperature, and, therefore, the expansion of the steam and its velocity at the moment of impact is also fixed; further, the water in the suction pipe is drawn into the tube only by the internal vacuum; the influence of the entrance area extends also to the velocity of the feed water, as it approaches in a thin sheet, the actuating steam. From the laws of impact, it is found that the greater the difference between the velocities, the greater will be the loss of actual energy at the moment of impact; as the whole transfer of the mechanical energy of the steam to the water is by impact, it follows that there is great advantage in giving to the entering water the highest possible velocity; this can only be obtained by maintaining the pressure in the combining tube as low as possible, and reducing the water entrance to a minimum. The lower this pressure the higher will be the velocity of the entering water, and the greater the proportion of water to steam in the mixture.

Secondly, the convergent taper or curve extending from the end of the steam nozzle to the lower overflow, should, in order to obtain the maximum efficiency from the injector, conform closely to the rate of condensation of the steam, and its length be modified for every variation in the pressure of the steam or the temperature of the feed. The same condition obtains to a great extent with the water entrance area, whose influence upon the performance of the injector was detailed in the preceding paragraph. It was appreciation of these facts that led Giffard and the early experimenters to lay so much stress upon the necessity for an adjustable combining tube. The advantage of this feature can be better understood if we suppose an injector to be working into a boiler under normal and efficient conditions, lifting the feed water one foot and running at its maximum capacity; if the steam pressure now rise, more water will be required to condense the increased flow of steam and preserve the normal condition of the jet; this can only be effected by increasing the head of water or widening the distance between the steam nozzle and the combining tube. A reduction of pressure would evidently require a reverse movement, for the vacuum within the tube remaining constant, too large a quantity of water would enter for the steam to force through the delivery tube, and waste would occur at the overflow. If the steam pressure fell much lower, the jet of steam would not have sufficient power to drive the accumulating mass of water through the lower end of the combining tube, and the continuity of the jet would be lost. If the openings in the tube, *i. e.*, the overflows, were large enough, all the water would pass out through the waste pipe instead of going into the boiler, even though the sound of working be the same as that with which the engineer might be familiar; on the other hand, if the overflows are small, the injector will "break," or "fly off," and steam and hot water will be forced down into the suction pipe.

The following experiments with the "Little Giant Injector" illustrate these conditions as they occur in practice; a No. 7 injector was started at 90 pounds steam on a lift of one foot, and the combining tube adjusted for the full capacity; the position of the tube was measured and found to be 15 millimetres from the upper end of its stroke. The steam pressure was then raised to 120 pounds and the temperature of the delivery increased from 136° to 165°; as this value was entirely too high for efficient performance, the tube was moved 10 mm. further down and the temperature fell at once to 148°.

Starting once more with 90 pounds steam and delivery at 136°, the height of the lift of the feed water was increased to 6 feet, the result being a diminution in the capacity of the injector and a delivery temperature of 150°; a downward movement of the combining tube of 10 mm., brought the

THE COMBINING TUBE.

capacity up to the standard, and reduced the delivery temperature to 136° again; the distance between the tubes under these conditions being the same as was used at 120 pounds when lifting the feed water 1 foot.

The following table gives the distance between the combining tube and the steam nozzle at different pressures, both for the maximum and the minimum capacities; the height of lift is 1 foot; it will be noticed how much variation in the position of the tube there is between high and low pressures.

Steam Pressure.	Distance of Combining tube from upper end of Stroke.	
	Max. Capacity.	Min. Capacity.
120	24 mm.	3.5 mm.
90	15 "	1.0 "
60	7 "	0.5 "
30	1.5 "	0.5 "

The wide difference between the position of the combining tube at 30 and at 120 pounds steam show how impossible it is for a single jet injector with fixed nozzles to work thoroughly efficiently over a large range of pressures. The only means by which this is attempted, is by assuming a certain range through which the pressure may fluctuate, and then adjusting the combining tube so that the injector will work into the boiler, without wasting at the overflow and at its maximum capacity, at the lower limit of pressure; this method sacrifices the best performance of the injector at higher steam pressures, as both the overflowing temperature and the capacity will be lower than if the tubes were differently adjusted. In the previous example, if the combining tube is set for a steam pressure of 40 pounds, the adjustment would be correct for the minimum capacity at 120 pounds; therefore, as the pressure was increased, the efficiency of the instrument would diminish, and a less proportion of water be delivered per pound of steam.

This demonstrates the superiority of the adjustable or self-adjusting form of injector over the fixed-nozzle type for use in all places where the steam pressure is subject to much fluctuation; the correct ratio between the weight of the water and the weight of the steam can always be maintained in the former case, and the capacity increased with the steam pressure; even though a throttling valve may be placed in the feed pipe, the results obtained where there is considerable range of pressure, as in locomotive service, cannot be as satisfactory as with the other method of regulation.

The form of the combining tube as determined by various experimenters differs greatly; in the tests of the Irwin Injector by a committee of the Franklin Institute, in 1879, a tube was used whose length was only four times the diameter of the delivery tube; this is a great contrast to that of the exhaust injector, where the ratio is 18 to 1; this corresponds to the usual practice for high pressure steam, where the increased quantity of heat requires provision for its absorption, and this can be best accomplished by lengthening the tube, and giving better opportunity for intimate mixture between the water and the steam; with the exhaust injector this is necessary on account of the large volume of steam used, which requires ample time for condensation, for the temperature of the delivery is higher than that of a well designed live steam injector working at 150 pounds.

The advantages of the adjustable combining tube for varying the water area are further increased if this adjustment be made automatic and effected by the action of the jet itself; that this has been done, was shown in the description of the self-adjusting injector, accompanying Fig. 3; in this case the automatic action was described as effected by the influence of the jet passing the overflow space, and it was shown that the combining tube was moved forward by the partial vacuum produced by incomplete condensation of the steam, or backward by the pressure due to excess of water in the feed chamber. It is interesting to note the action of the steam upon the upper end, which is as follows: the discharging jet, striking against the film of water on the inside of the tube, tends to impel it forward, while the rapid condensation of the steam produces a reactive effect which draws the tube backward toward the steam nozzle, and these two pressures almost exactly balance; but as this action only

THE COMBINING TUBE.

occurs in the central conical part, the rest of the piston head receives the positive or negative pressure in the feed pipe, just as the lower end is acted upon by the pressure in the confined overflow chamber; this has been found to be true by placing a vacuum gauge upon the overflow, and noting its reading as the height of the lift increases; it is found that the two readings correspond almost exactly, and that the tube floats between two balancing cushions, ready to respond to any change in the governing conditions.

In calculating the performance of an injector, it is often desirable to know the vacuum within the combining tube. This may be determined by allowing water under a constant head to flow freely through the tube, special care being taken to see that the feed valve is full open and that the upper overflow is large enough to permit free discharge for all the water that will enter; the best proof of this is to disconnect the steam branch and observe if the water rises into the steam nozzle; if it does, the overflow space is too small to permit a free discharge. If, however, this test is satisfactory, weigh the quantity of water flowing through, and then connect the steam pipe and admit steam; without altering the position of the water valve, regulate the supply of steam until the injector will just run without wasting, and note the new weight of water; these two values will bear the same relation to each other as the square roots of the heads.

Here is an actual test of an exhaust injector, taken under rather unfavorable conditions as the feed water was 76 degrees: under a constant pressure of 4.25 pounds or 9.817 feet, 1987.5 pounds of water flowed through the injector; with the steam valve opened and working into boiler, the weight increased to 3887.5 pounds, due to the vacuum in the combining tube Therefore,

$$1987.5 : 3887.5 :: \sqrt{9.817} : \sqrt{37.48} = \text{total head}.$$

Subtracting,

$$37.48 - 9.817 = 27.663 \text{ feet vacuum}, = 24\frac{1}{2}'' \text{ mercury},$$

which is the value required.

The average vacuum within the tube cannot vary much from that corresponding to the mean temperature of the feed and the delivery, yet the pressure at the point where the water enters is much lower; it is there that the water is the coldest, and the outside edges of the discharging jet obtains its fullest expansion. This will be shown in the diagram showing the discharge from the steam nozzle (see page 54), and described under that heading. It is interesting to determine as closely as possible the average pressure within the tube, for it is only by this means that the final expansion of the steam can be found, and its terminal velocity calculated.

So far it has only been assumed that there was a partial vacuum in the tube. As it is obvious that steam is always condensed when coming in contact with a body whose temperature is lower than that corresponding to its pressure, so it is possible that steam may be condensed in an injector tube even when the entering water is above 212 degrees. This is most nearly approached in the double jet injector, where the water entering the combining tube of the second set of tubes is often as high as 180 or 190 degrees, when the temperature of the supply is 150. It is very probable that in this case only a very small percentage of the steam from the second steam nozzle is condensed, but sufficient to permit the combined jet of steam and water to contract sufficiently to pass through the narrowest section of the delivery tube. The velocity with which the water enters the tube is also much increased, for the pressure between the two sets of tubes rises—at 120 pounds steam—to 40 or 45 pounds; so that the work required of the second steam jet is not very great.

There is no doubt that if an indefinite amount of time were given for the condensation of the steam, feed water could be taken at any temperature below that of the steam, but it must be remembered that the actual time of contact is only that required to traverse the tube, and that the mixing can only be between the two conical exposed surfaces; therefore, if the difference in temperature be small, the transmission of heat will be correspondingly slow, and the volume of the steam will not be reduced enough to enter the delivery tube. The effect would be similar to that of an air jet discharging into a mass of water—a dispersion and atomizing of the whole mass. It is evident, therefore, that the thinner the sheet of entering water and the lower the final velocity of the jet, the greater will be the efficiency of the injector.

For each special purpose the tube has to be designed, and the form best adapted for high temperature is not that which is best suited for low pressures of steam or for high lifts, so that the shape adopted for ordinary practical results is usually the mean for the conditions under which the injector is to operate, regarding which each designer is apt to have his own opinion.

Temperatures as high as 162 degrees, and variations of capacity up to 75 per cent. have been obtained in experimental work by forms of the Sellers' Fixed Nozzle Injector with varied forms of tube, neither of which the writer believes has ever been exceeded by any other styles of injector with which he is familiar.

The advantage of maintaining the feed water pure and free from lime and dirt is more apparent in the case of the combining tube than with any other part of the injector; as the specific gravity of sand and grit is greater than that of the water, all foreign particles are driven with great force against the walls of the tube by the impact of the steam. The tendency is not only to enlarge the diameters at different points, but also to change the shape by wearing and grinding off all shoulders against which the jet can strike; soft spots in the brass, caused by unequal mixing of the metal before casting, soon show the effect of abrasion, and this introduces a diagonal motion to the particles that wear depressions upon the opposite side. Tubes that have been used with impure water show this feature very strongly, and the conical shape is frequently changed to that of two cylinders, connected by an abrupt shoulder. The force of the impact of the jet of steam is so severe, that a tube with walls $\frac{3}{16}''$ thick was found to have a pocket $\frac{1}{8}''$ deep worn in it at the end of the steam nozzle, and the outside was considerably bulged by the continuous blows it had received.

But with pure water, a well designed tube will last a long time, and in railroad service, where so many injectors are employed, the choice of the water supply should take into consideration not only the chemical analysis, but also the percentage of impurities mechanically mixed with it; in many cases these impurities are almost invisible to the naked eye, and it is only by careful filtration that the trouble can be remedied. On many of the railroads of France, this system of purification is carried out very perfectly, and the effect is apparent not only in a reduction in the cost of repair to the locomotive boilers, but also in the longer life of the injectors.

CHAPTER VI.

THE STEAM NOZZLE.

THE advantage of obtaining the highest velocity of the steam at the instant it strikes the water has already been shown, so that the great importance of obtaining the best shape for the nozzle which guides the discharge of the actuating jet, is at once apparent.

As it was assumed in the earliest experiments that the discharge of gases followed the same laws that govern inelastic fluids, the steam nozzles of the first injectors were made with a gradual convergent taper, and it was not until 1869 that the form was improved by the application of a divergent flare, that permitted expansion of the steam within the limits of the tube. The effect of this change upon inelastic fluids, like water, would have been to increase the volume discharged and diminish the terminal velocity; but with steam, air or other gases, the gradual lowering of the pressure as the jet traverses the tube, gives an increase in volume and expansion in the direction of the flow, which augments the velocity of the particles of the fluid, while the weight discharged remains unchanged.

In order to demonstrate more clearly the great difference between the discharge of gases and liquids, some experiments will be described that were made with tubes of different forms, using steam under pressures of 120 and 60 pounds (gauge). Sketches and instantaneous photographs were taken of jets of steam discharging into the air, and the form of the jet and the direction of the motion of the particles were carefully noted. The results are given in Fig. 15, which accurately represents the external form assumed by the jets, although it does not show the change from transparency to whiteness that occurs shortly after the steam leaves the nozzle. Four styles of tubes are given:

1. A convergent nozzle or short cylinder.
2. An aperture in a thin plate.
3. A divergent tube, straight taper.
4. A divergent tube, curved taper.

FIG. 15.

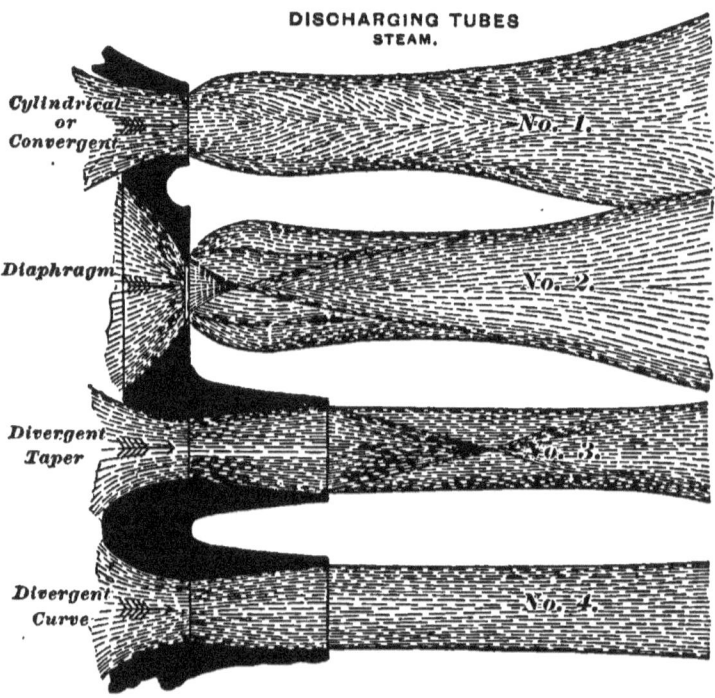

The first tube represents the shape that would be used to produce a solid water jet at a high velocity, and is similar to the earliest form of the steam nozzle. The second is a thin diaphragm, or any orifice where the thickness of the walls is small compared with the width of the opening. The third is shaped like a divergent cone, widening in the direction of

THE STEAM NOZZLE. 55

the flow. The fourth has the same dimensions, but expands in curved lines from the throat to the lower end.

In all these experiments, the jet, after leaving the end of the nozzle, was almost invisible for a distance of two or three diameters, and of a pale bluish color, marked with light lines of white, apparently produced by the entrained particles of water. Beyond the transparent portion, the jet expanded to a much larger diameter, became white on the surface, and was finally condensed by the cooling effect of the air.

The difficulty with the first two tubes is that they permit immediate diametral expansion of the steam, instead of confining it and compelling expansion in the direction of motion. A transparent envelope is formed, three or four times the size of the orifice, through which the central jet, discharging at somewhat higher velocity, can be distinctly seen. This swelling of the jet is due to the fact that the internal pressure at the moment of discharge is greater than that of the medium into which it is flowing; it is most apparent in No. 2, because in this tube the internal pressure of the steam at the terminal section is the highest, and in all cases is more marked with high than with low steam. In the divergent nozzle, the terminal pressure is the same as that of the atmosphere; if it were higher, there would be the same enlargement that occurs in the previous cases, and if lower, a contraction would be caused by the pressure of the air. In these two nozzles, the direction of discharge of the particles is almost exactly parallel with the axis, and therefore, all the energy of the steam, except the slight loss due to friction against the walls, is utilized for augmenting the velocity.

To find the velocity of discharge a knowledge of the internal condition of the jet is essential; during the movement of the steam toward, and through the tube, there must be a reduction in its pressure, and a corresponding increase in volume. If then, the area of the nozzle at different sections is known, it will be necessary to find the pressure at those points in order to determine the volume or density from which the velocity may be calculated. This was done when the injector was forcing water into the boiler, and also when the steam was discharging freely into the air, by inserting a small tube along the axis of the nozzle, and observing the indications of a gauge placed on its outer end. Communication was made with the interior of the jet by means of a small hole drilled through the tube, so that by sliding the tube backward or forward, the internal pressure of the jet could be obtained within or beyond the limits of the nozzle, either when the injector was working, or during free discharge. By this means the pressure within the transparent portion of the jet was ascertained, and the fall of pressure during condensation. The experiment was also tried of drilling minute holes through the steam nozzle normal to the jet, and reading the gauge directly, but by this method observations were restricted to the limits of the nozzle and could not be obtained when the injector was in action.

The results of these investigations are shown in Fig. 16, where A is the steam nozzle, B the combining tube, f the hollow spindle with transverse hole f', by which the pressure was communicated to the gauge. The intersections of the vertical lines with the curved lines in the diagram indicate the pressure at the respective points of the nozzles; the horizontal lines represent pressure above or below the atmospheric line. Observations were made at 60 and 120 pounds gauge pressure, the full line showing the conditions when the injector is running, and the dotted line when the steam nozzle is discharging into the air.

Considering first the case where the injector is working, it will be noticed that as the particles of steam approach the entrance to the nozzle, the pressure falls in easy curves until the smallest part of the tube is reached, when the descent is more abrupt, but approaching the usual form of expansion as shown by the indicator card of a steam cylinder. Just beyond the end of the steam nozzle, at the line $a'a'$, the proximity of the feed water causes a quick fall of pressure that is only partially recovered during its passage through the combining tube.

THE STEAM NOZZLE. 57

These pressures of 22" for 120 pounds, and 24" for 60 pounds, are found at the centre of the jet; that due to the actual contact of the feed water with the steam envelope would approach more nearly a perfect vacuum.

FIG. 16.

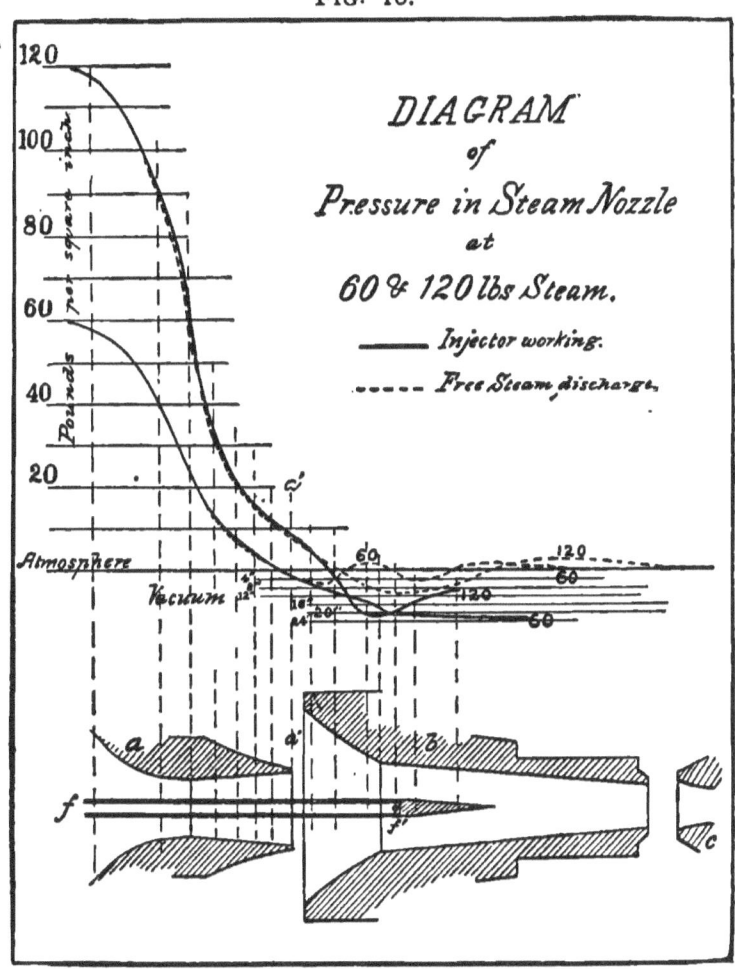

Turning now to the freely discharging steam, it will be observed that during passage through the nozzle, the lines for the two conditions almost overlie, and probably would, if all slight errors of observation could be excluded; this seems singular when it is considered that in one case the steam is discharging into a partial vacuum, and in the other, into the air; beyond the line $a'a'$ the curves separate, yet both pass below the atmospheric line, the curve of free discharge rising and falling, owing to the unstable equilibrium of the jet. Under 120 pounds initial pressure the jet emerging from the tube at 9 pounds pressure, expands to atmospheric, and then by internal condensation, 11" vacuum is reached, but is soon overcome and equilibrium established. This peculiar phenomena of alternate rising and falling of the internal pressure is thoroughly borne out by the external appearance of the jet, as it presents the appearance of having nodes separated by swelling curves. The 60 pound pressure line shows that the steam leaves the nozzle at a pressure of 4" vacuum, and, therefore, contracts instead of showing the surrounding envelope, characteristic of the other case.

From the correspondence of the curves showing the fall of pressure, under both conditions of discharge, it appears that the outflow is unaffected by the pressure in the receiving chamber; when the injector was working at 120 pounds, the pressure beyond the end of the steam nozzle was 22" vacuum, and in the other case, 14.7 pounds, yet the conditions within the limits of the nozzle were almost precisely the same. If the experiments were carried still further, and the weight of steam determined that would pass in a unit of time through an orifice in a reservoir in which the initial pressure was maintained constant, while the pressure in the receiving tank was increased from a vacuum up to the upper limit of pressure, it would be found that the rate of flow would be practically constant until the counter pressure was raised to $\frac{6}{10}$ of the initial. It is true theoretically, and has been also proved by careful experiment, that this value of $\frac{6}{10}$, corresponds closely to the relation of pressures that will give *maximum* flow, although the actual increase above the weight discharged against atmospheric pressure is small. At 120 pounds gauge pressure, for example, a little more steam will be discharged against 66 pounds than there would be into the air, for, taking the absolute pressures to which this ratio applies, $(120 + 15) \times 0.6 = 81.$ = the absolute, or $81 - 15 = 66.0$ = the gauge

FIG. 17.

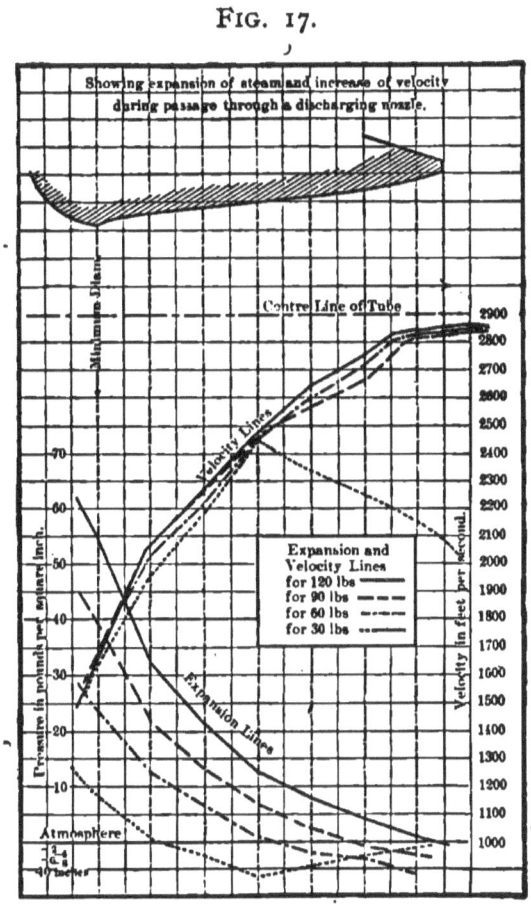

pressure. As an explanation of this peculiarity of the jet, Rankine has suggested that there is always a limiting section at which the internal pressure is 0.58 of the initial, and until that value is exceeded at a lower part of the jet, the flow will not be reduced.

With apertures in thin plates, like No. 2, Fig. 15, or short cylindrical tubes, the limiting section is found directly at the opening; but with many of the steam nozzles as designed for injector use, it is a little in the rear of the minimum diameter; this is true even when the tube may be designed to permit immediate free expansion, as shown in Fig. 17 where, although the pressure at the smallest diameter is below 0.58 of the initial, the velocity at that point is nearly constant. The diagram shows a half section of a nozzle divided by vertical dotted lines that pass through minute holes that were drilled for the purpose of obtaining the pressure of the jet by means of an ordinary gauge. Readings were taken at the different steam pressures, and the results plotted under the respective parts of the tube. The pressure lines are similar to those in Fig. 16, and indicate the rate of expansion of the steam while traversing the nozzle. The upper curves are for the velocity, and are calculated from the observed pressures; these curves almost overlie throughout their whole length, and probably would, if the flare of the nozzle were still further expanded.

The throat pressures, ratios and velocities are as follows:

Initial Pressure, Absolute.	Throat Pressure, Absolute.	Ratio.	Velocity at Throat. Feet per Sec.
135	69.8	0.517	1612.0
45	22.8	0.508	1603.0

Although there is a difference of 95 pounds between the extreme initial pressures, there is only a difference of 9 feet in the two velocities. It is possible that the orifice by which the pressure was recorded on the gauge in these experiments, was a little below the limiting section—probably not more than $\frac{1}{100}$ inch—as the smallest change in its position gives a wide variation in the pressures.

The ordinary type of injector steam nozzle gives slightly different results, for the divergent taper is straight and does not allow as great an amount of transverse expansion, so that the ratio of the throat pressure to the initial pressure is not constant. With a tube having a divergent flare of 1 in 6 in diameter, the following results were obtained:

THE STEAM NOZZLE.

Absolute Pressure.		Ratio.	Velocity in Throat.
Initial.	Throat.		
135	82.0	0.606	1407
105	61.5	.585	1448
75	42	.559	1491
45	24.5	.546	1504

In this case, the nozzle was formed with easy curves of approach to a short cylindrical portion whose length was 0.3 the diameter and into which the gauge hole was drilled; from this point the tube widened with a taper of 1 in 6. This tube corresponded to a form of steam nozzle extensively used in injector service, and proves, from the fact that neither the pressure ratios nor the velocities are constant, that the actual rate of expansion is not similar to that of free discharge, nor that of the widely flared nozzle shown in Fig. 17.

If the tube is cylindrical the weight of steam flowing per second reduces with an increase in length; at 75 pounds the flow through a ½" tube was,

½" long 900 lbs. per hour.
1" long 892 lbs. per hour.
1½" long 864 lbs. per hour.

To determine the weight of steam passing through an orifice, it is often convenient to use a simple formula instead of resorting to actual measurement. The following equation was found by R. D. Napier to give results very close to actual weighings, with results only about 2 per cent. low at 70 pounds, and 1 per cent. low at 120 pounds; Rankine has also recommended it for approximate calculations. The formula is

$$w = \frac{P \times A}{70} \quad \ldots \ldots \ldots \ldots \ldots (9)$$

where
$A =$ area of the orifice in square inches,
$P =$ absolute pressure, $=$ gauge pressure plus 15,
$w =$ weight of steam discharged in pounds per second.

Knowing the weight of a cubic foot of steam at the pressure in the orifice, the velocity of the jet at that point can easily be found. It is better, however, to apply the simple approximate formula given by Rankine, which is based upon the energy exerted by the steam during expansion:

$$V = \sqrt{2g\,U} = 8.025\sqrt{U} \quad \ldots \ldots \ldots \ldots (10)$$

$$U = 778\left[\frac{(T_1 - T_2)^2}{T_1 + T_2}\right] + \left(\frac{T_1 - T_2}{T_1}\right)(1105590. - 510.2\,T_1)\ldots(11)$$

Here V is the velocity in feet per second at the lower pressure where the absolute temperature is $T_2 = t + 461.2$; and T_1 the absolute initial, which at 120 pounds steam is $T_1 = 350.03 + 461.2 = 811.23$ degrees; the pressure being known, the actual temperature of the steam can be obtained from any steam table, and the absolute temperature, by adding 461.2. The use of this formula is to calculate the velocity of the jet at the instant the particles of steam strike the water and the transfer of momentum is effected; this one is the simplest to apply, as it is not complicated by other calculations.

The application of this formula for finding the velocity requires the assumption that the work performed by steam during discharge is equal to that exerted against a piston moving in a cylinder, and that no heat is received from, or given out to, external bodies; in other words, the relation between pressure and volume in all parts of the jet follows the law of adiabatic expansion. During this process, a certain percentage of steam is condensed, which increases with the difference between the initial and terminal pressures; for example, in expanding from 120 pounds to the atmosphere, $12\frac{4}{10}$ per cent. of the weight discharged is condensed, leaving $87\frac{6}{10}$ per cent. of gaseous steam; when the final pressure is—22" vacuum, as in the case of the injector quoted on page 58, 1 pound of steam at the moment of impact will contain 0.179 pound water and 0.821 pound steam.

These values may be calculated for any degree of expansion by means of formula (12):

$$x_2 = \frac{\theta_1 - \theta_2 + \frac{x_1 r_1}{T_1}}{\frac{r_2}{T_2}} \quad \ldots \ldots \ldots \ldots (12)$$

THE STEAM NOZZLE. 63

where x_1 and x_2 are the weights of steam at the absolute initial and terminal pressures. (If initial steam dry, $x_1 = 1$.)

T_1 and T_2, = absolute temperatures = $(t + 461.2)$,
r_1 and r_2, = latent heat under the two conditions,
θ_1 and θ_2, = entropy of liquid.

These terms can be obtained from Steam Tables; the values of θ most frequently used are, however, given in Table III, page 66.

Formula (No. 12) will be found of use in calculating results from experimental data, and for supplying the value of x_2 in (13), by which the velocity of the jet can be conveniently found:

$$V = 8.025 \sqrt{778\,(x_1 r_1 + q_1 - x_2 r_2 - q_2)} \quad \ldots \ldots \ldots (13)$$

and the weight discharged,

$$w = \frac{A\,V}{144\,(x_2\,(S - 0.016) + 0.016)}. \quad \ldots \ldots \ldots \ldots (14)$$

where
V = velocity.
w = weight discharged, in pounds per second.
A = area of the nozzle at smallest section, in square inches.
q_1 and q_2 = heat of the liquid at the terminal pressures.
S = volume 1 pound steam in cubic feet, at final pressure.

By way of illustrating the use of these formulæ, the velocity of a jet of steam discharging from a reservoir at 120 pounds pressure into the air will be calculated. The steam will be supposed to be dry, and to follow the law of adiabatic expansion. The value of x_2 must be first determined by (12). As the steam is supposed to be dry in its initial state, $x_1 = 1$; the other terms are as follows:

$\theta_1 = 0.5027$ $r_1 = 867.3$ $T_1 = 350.3 + 461.2 = 811.5$ $q_1 = 321.4$
$\theta_2 = 0.3135$ $r_2 = 966.07$ $T_2 = 212 + 461.2 = 673.2$ $q_2 = 180.5$

$$x_2 = \frac{0.5027 - 0.3135 + \dfrac{867.3}{811.5}}{\dfrac{996.07}{673.2}} = 0.876$$

$V = 8.025 \sqrt{778\,(867.3 + 321.4 - 0.876 \times 966.07 - 180.5)} = 2848.2$

Formula (10) gives 2827 feet per second, which agrees fairly well with this result.

If dry steam at the same pressure expand to $-22''$ vacuum, or 4 pounds absolute, the value of x_2 will be found to be 0.8243, and the final velocity, 3446 feet, or ⅔ mile per second; calculated by (10), V is 3443 feet.

The presence of water in the steam is shown by the change in the appearance of a jet discharging into the air; the clear bluish portion of the jet adjoining the orifice becomes an opaque white. The effect upon the velocity can be found by giving a lower value than unity to x_1 in formula (12). If 5 per cent. of water is entrained, $x_1 = 0.95$, and in the previous example x_2 becomes 0.7899, and the final velocity is 3223 instead of 3446 feet.

Besides retarding the velocity, entrained water has an injurious effect upon the surfaces of the tubes, cutting and abrading the metal wherever soft spots occur; dry pipes should be used to prevent this trouble and all pockets or depressions in the steam pipe avoided; for prompt starting, efficient action, and general reliability of an injector are much enhanced by attention to this apparently trivial condition.

The area of the steam nozzle, upon which depends the weight of steam used per hour, can best be defined in terms of the area of the delivery tube. This varies with the pattern of the injector, and the purpose for which the instrument is designed. From superficial considerations, it would appear that if this ratio were made unity,—*i. e.*, the steam nozzle area = delivery tube area,—the jet would just have sufficient power to sustain the initial pressure; the nearest approach to this in experimental work is with the ratio 1.007 to 1.000, but in actual practice, on account of the necessity of a margin of counter pressure, the ratio varies from 2 to 1, to 3 to 1; in the exhaust injector this is increased to 16 to 1, as the ratio of the initial to the terminal pressure is 1 to 6, or 16 pounds to 96 (absolute).

The motive element in the injector is the steam supply, and the work performed is raising the water to the level of the instrument and delivering it against the pressure of the boiler. From the stand point of mechanical efficiency, it is obvious that the smaller the weight of steam required to perform this work, the more economical will be the operation. The condensation of the motive jet is the source of economy in this method of feeding, as the only loss is the small amount heat radiated. The employment of a larger quantity of steam than is absolutely necessary, would be using the instrument as a heater, or similar to the application of a jet of steam to the suction pipe of a pump for the purpose of warming the water.

The raising of the temperature of the feed water is beneficial, but can best be accomplished by other means; by utilizing the heat of the waste products in the smoke-box or stack, or by using the heating surface of the boiler itself, instead of absorbing available energy from the boiler for the purpose of performing subordinate work. This is specially applicable to service on locomotives, where the high rate of speed employed at the present time necessitates the strictest economies in the use of steam, and the application of the full capacity of the boiler to its primary use, supplying steam to the cylinders; it frequently happens that the injector can only be placed in operation when the excessive strain is relieved by a stop at a station or during a run on a down grade. On the other hand, the most economical results are obtained, not by an intermittent water supply, but by maintaining a constant water level and a continuous feed.

The actual amount of steam required during the feeding of a locomotive boiler is not generally realized; an evaporation of 2,500 gallons of water per hour, requires, with some patterns of injectors, 2,300 lbs. of steam, which, at an estimated rate of 35 lbs. per H. P., amounts to 65 H. P. If the steam were used to best advantage, this could be reduced to about 45 H. P., and the temperature of the water entering the boiler would be lowered approximately 25°. It is of course advantageous to supply the boiler with water as hot as possible, to avoid shrinkage and unequal expansion of the sheets; but, as can be easily seen, the benefits gained in that direction by raising the temperature of the feed by means of a jet of high pressure steam, seldom compensate for the loss of available steam supply.

There can be no question but that the injector is superior to all other devices for feeding locomotive boilers that have yet been introduced, but its action should be as efficient as possible, in order that the steam and fuel consumption which may properly be charged to its account, can be reduced to the minimum.

A short table giving the value of θ may be found to be convenient when complete tables are not at hand:

TABLE III.

Absolute pressure, lbs. per sq. in.	θ	Absolute pressure, lbs. per sq. in.	θ	Absolute pressure, lbs. per sq. in.	θ	Absolute pressure, lbs. per sq. in.	θ	Absolute pressure, lbs. per sq. in.	θ	Absolute pressure, lbs. per sq. in.	θ
1	.1329	6	.2480	14.7	.3135	40	.3921	65	.4337	90	.4633
2	.1754	7	.2587	20	.3363	45	.4020	70	.4402	100	.4733
3	.2013	8	.2682	25	.3539	50	.4109	75	.4464	120	.4911
4	.2203	10	.2842	30	.3685	55	.4191	80	.4522	150	.5133
5	.2353	12	.2976	35	.3811	60	.4267	85	.4579	200	.5429

An efficiency comparison of different styles of injectors can be made either by rating the delivery of the injector in cubic feet or pounds per hour in terms of the area of steam nozzle, or by determining the ratio of the weight of steam used to the weight of water forced into the boiler. In this connection may be given the following tests of some of the best-known patterns of locomotive injectors, all made under precisely the same conditions: steam 120 pounds, feed temperature 65° Fahr., height of lift 18″. Steam dry.

Name of Injector.	Nominal Size.	Weight of Water Delivered per lb. of Steam.
Belfield	No. 10	9.69 pounds
Garfield	No. 7	13.53 "
Little Giant	No. 7	12.92 "
Mack	No. 7	13.79 "
Metropolitan	No. 9	13.16 "
Monitor	No. 9	11.31 "
Sellers' 1887	No. 8½	13.80 "

CHAPTER VII.

THE ACTION OF THE INJECTOR.

NOTWITHSTANDING the fact that much has been written upon this subject, the action of the injector still appears mysterious to many of those who are familiar with its operations. It is strange that the reason for its working is not more generally understood, even by those accustomed to operate it daily, especially as this method of feeding is now so universally employed for locomotive and stationary boilers.

The simplest method of considering the theory of the injector is to eliminate the more complicated sides of the question and consider it solely from a mechanical point of view; simply as an apparatus in which the momentum of a jet of steam is transferred to a more slowly-moving body of water, producing a resultant velocity sufficient to overcome the pressure of the boiler.

The high velocity attained by a jet of steam has been calculated, and diagrams have been given that show the fall of pressure and increase in velocity as the volume is increased according to the laws under which the steam expands. Suppose that a nozzle connected with a reservoir containing steam at 120 lbs. pressure discharges 1 lb. of steam per second; at its minimum diameter, the steam will have reached a velocity of 1407 feet, but when the terminal pressure is 22" vacuum, the velocity will be 3446 feet per second. Let us suppose that this jet flows into a combining tube, which is able, by means of the great conductivity of its walls, to abstract sufficient heat to completely condense the steam at a final pressure of 22", or 4 lbs. absolute. This reduces the steam to a solid jet of water having a cross section $\frac{1}{714}$ the area of the steam while passing through the steam nozzle, and yet does not in any way affect the velocity, as the contraction of the jet is entirely lateral. A jet of water issuing from the delivery tube, forced out by the pressure of the boiler, would have a velocity nearly equal to that due to the head, or approximately, 133 feet per second, only $\frac{1}{25}$ of that of the jet of condensed steam; but an injector is required to perform useful work, forcing a supply of feed water into the boiler; therefore a certain weight of feed-water must be added which will take the place of the cold walls of the tube for the purpose of condensation. This mass of water receives the energy of the moving steam, condenses it, and the two fluids move along together through the delivery tube with a terminal velocity greater than a jet of the same density issuing *from* the boiler. If the weight of water supplied is too great, the steam will not have power enough to give the required velocity of 133 feet; if there is an insufficient supply, the volume of the steam will not be reduced sufficiently to pass through the tubes, and in neither case will the injector work properly.

Turning again to figures, and taking the simplest possible case, we can follow the steam through its whole course within the tubes, and determine the relation of the different parts of the injector. As the velocity and volume of the steam at the instant of passing the minimum diameter of the steam nozzle are 1407 feet, and 5.05 cubic feet, respectively, and after complete expansion in the combining tube 3446 feet (see page 64) and 74.2, the cross section of the steam jet at that time must be,

$$\frac{1407}{3446} \times \frac{74.2}{5.05} = 5.99 \text{ times the area of the steam nozzle.}$$

During complete condensation, the volume of 1 lb. of steam shrinks from 74.2 cubic feet to 0.16 cubic feet, so that the cross section of the jet after condensation is

$$\frac{0.016}{74.2} \times \frac{5.99}{1} = \frac{1}{774} \text{ the area of the steam nozzle,}$$

so that the jet of condensed steam would pass through an

THE ACTION OF THE INJECTOR. 69

orifice $\frac{1}{714}$ the area of the steam nozzle at a velocity of 3446 feet per second.

The substitution of a cone of water as the condensing medium, at a ratio of 13 pounds of water to the pound of steam—as this is a fair performance of a locomotive injector at this pressure—will increase the volume of the jet (13 + 1) = 14 times, and require the area of the delivery tube to be enlarged to $\frac{14}{714}$. Further, the addition of this weight of water will retard the motion of the jet, and demands a still larger orifice of entrance; it will therefore be necessary to determine the final velocity of the mixture before the size of the delivery tube can be obtained. The transfer of the energy of the steam to the water is similar to that occurring during the impact of two inelastic bodies, and owing to the fact that the particles of steam do not all strike the molecules of water in the proper direction, and owing to the obliquity of the entrance of the feed into the combining tube, and also other causes, a loss occurs which varies somewhat under different conditions, but under those assumed, amounts to 40 per cent. Multiplying the weight of the water by its entrance velocity of 40 feet due to the 22″ vacuum in the combining tube, and adding the momentum of 1 lb. of steam at a velocity of 3446 feet, we have the following equation of momentum, which determines the velocity of the mixture:

$$(1.00 - 0.40) \times (40 \times 13 + 3446 \times 1) = 14 \times v$$

$$v = 169.97 \text{ feet per second.}$$

The temperature of this mixture would probably be about 150°, and 1 lb. per square inch, from Table II, would be equal to a head of 2.355 feet, and from $h = \dfrac{v^2}{2g}$, and $h = p \times 2.355$ we have

$$p = \dfrac{v^2}{2.355 \times 64.4} = \dfrac{28889.}{151.66} = 190.5 \text{ pounds back pressure}$$

corresponding to this velocity; but from this must be subtracted the pressure in the combining tube below the atmosphere, 22" or 11 lbs., so that the available counter pressure as shown on the gauge will be 179.5 lbs.

The velocity of the original jet is now reduced from 3446 feet, as in the assumed case when passing through the delivery tube, to 169.97 feet, requiring a further enlargement of the area, in order that the augmented volume of the jet may find entrance at this reduced velocity. Summing up the two changes, one due to the decrease in velocity, and the other due to increase of volume, we have,

$$\frac{3446}{169.97} \times \frac{14}{774} = \frac{1}{2.72}$$ the area of the steam nozzle;

or, in other words, if the area of the delivery tube be taken as unity, the area of the steam orifice will be 2.72, and the ratio of the diameters will be 1 to 1.65, which approaches closely the proportions in ordinary practice.

It is thus seen that the whole action of the injector depends upon the fact that the velocity of a jet of steam discharging into the combining tube, is 20 to 25 times that of a jet of water issuing from a boiler under the same pressure, and that the enormous reduction of the volume during condensation concentrates the momentum of the jet upon an area which is but a small fractional part of the orifice from which it issues, leaving a large margin of available energy which may be applied to useful purposes. As condensation plays such an important part in the operation, it is seen that any condensible gas may be substituted for the motive steam, if the inherent conditions are properly considered, but some modifications of the proportions of the parts as used in the steam injector might be found necessary in order to work satisfactorily under the new conditions.

The principle of the action of the injector working at 120 pounds steam into a boiler carrying the same pressure, may appear more easy of explanation than the case of an exhaust injector forcing water into a boiler at 80 pounds steam; that there is no difference beyond a change in the proportions of the parts, can be seen from the following example, which will be worked out according to the same analysis that was applied to the case of high pressure steam.

THE ACTION OF THE INJECTOR. 71

Assume the steam at 0 (gauge), or 14.7 pounds absolute pressure, containing, as it arrives direct from the cylinder, about 10 per cent. of moisture. From page 49 the vacuum in the combining tube will be assumed to be 24" or 3 pounds ab. From equation (12), we find that 1 pound of steam in its final condition will contain 0.832 pound of steam, and 0.168 pound of water; its velocity after complete expansion will be found from (13),

$$V = 8.025 \sqrt{778 \,(0.90 \times 966.07 + 181.6 - 0.832 \times 1015.3 - 109.8)} = 2205 \text{ ft.}$$

To work against a pressure of 80 pounds, the delivered water must be forced out of the combining tube, in which the pressure is 12.7 pounds below the atmosphere, so that the total pressure against which the jet must be capable of working is $80 + 12.7 = 92.7$ pounds; this requires a terminal velocity of 135 feet, under the assumption that the steam is all condensed and that its density is unity. The feed water enters under a head of 6 feet, and with a velocity corresponding to the sum of the partial vacuum and the head, approximately 48 feet per second. A somewhat greater percentage of loss must be taken than in the other case, as the action of the jet is not as efficient, and it does not seem as if it could be made so; taking 0.50 as the value of this coefficient, we have for the equation of momentum, from which the weight of feed water may be obtained,

$$(1 \times 2205 + 48 \times W) \times 0.50 = (W + 1) \times 135,$$

whence

$W = 8.74$ pounds of water per pound of steam,

which corresponds to the usual practice.

In both the examples given the simplest conditions were assumed and all uncertain elements avoided. The jet as it passed through the delivery tube was supposed to have the same density as water at the same temperature; in fact, this seldom occurs, as there is almost always a part of the steam uncondensed until the mixture has passed far into the delivery tube, and there is, in addition, a volume of air mixed with the steam that displaces a corresponding volume of water and gives to the jet its white, opaque appearance.

This quantity of air will depend upon the manner in which the boiler is fed, and the condition of the suction pipes and valves and stuffing-boxes of the injector or pump, as the total amount of air contained in the steam is greater than can be held in solution by the feed water. Experiments made with an injector placed entirely under water, so as to collect all the air discharged with the delivery, showed that this amount, though very variable, was by no means inappreciable; a No. 8 injector at 120 pounds, discharged 5.51 cubic feet of air per hour, and at 60 pounds, 4.38 cubic feet, measured at atmospheric pressure. These results do not represent the maximum quantity, but the mean of several tests.

The density, however, depends chiefly upon the percentage of steam condensed, and is, therefore, intimately connected with the water ratio. The following table shows the variation, and the increase in the velocity of the jet as the density decreases: steam pressure, 135 pounds absolute.

TABLE IV.

	Delivery Temperature. Degrees Fahr.	Weight of feed water per pound of steam.	Velocity of steam at time of impact. Feet per second.	Velocity of a solid jet of water at temp. t_2. Feet per second.	Total permissible counter pressure. Pounds per sq. in.	Actual velocity. Feet per second.	Density.	Coefficient of Impact.
	t_2	W	V	v	P	v_1	Δ	K
(a)	154°	12.00	3435.	165.3	180.	186.1	0.788	0.551
(b)	167°	10.33	3358.	174.0	199.	236.4	0.542	0.524
(c)	185°	8.63	3228.	181.6	215.	302.8	0.360	0.493
(d)	209°	7.02	3058.	177.3	202.	346.3	0.261	0.432
(e)	240°	5.60	2970.	147.2	143.	294.0	0.250	0.310
(f)	268°	4.69	2890.	141.6	130.	305.0	0.211	0.265

The ratio of water to steam that produces the highest back pressure is shown in experiment (c) as the energy of the jet is then at its maximum; with less water, the density of the jet is too low, even though the velocity is greater, and below that ratio the density rises, but the

THE ACTION OF THE INJECTOR. 73

velocity is insufficient, on account of the large increase in the weight of the mixture.

Simple formulæ can be applied to find the density: if the back pressure is known, find the velocity of a jet of water at the temperature of delivery, which designate by v, and the actual velocity by v_1. Let Δ be the density, and H the back pressure expressed in feet.

Then

$$H = \frac{v^2}{2g} = \frac{v_1^2 \Delta}{2g} \text{ whence } v = v_1 \sqrt{\Delta} \quad \ldots \ldots \ldots (15)$$

If A = the area of the delivery tube, the theoretical capacity $= v \times A$, and we obtain the following, by substituting (15) and reducing

$$\frac{62.4 \, A \, v_1 \, \Delta}{62.4 \, A \, v} = \frac{\text{Weight of actual jet}}{\text{Weight of jet unit density}} = \frac{\text{Actual capacity}}{\text{Theoret. capac.}} = \sqrt{\Delta} \quad (16)$$

so that

$$\Delta = \left(\frac{\text{Actual capacity}}{\text{Theoretical capacity}} \right)^2 \ldots \ldots \ldots (16)$$

The actual velocity in the table was found by dividing the calculated velocity by the square root of the density, and the changes in the velocity of the entering water and the discharging steam are due to the variation of the pressure within the combining tube with the temperature of the delivery.

Experiments (e) and (f) were made with the overflow closed, permitting a high pressure in the confined overflow chamber, otherwise there would have been a discharge of steam; but in many cases where two or more apertures in the combining tube are contained in the same chamber and only closed to admission of air by a light check valve, the delivery temperature may rise above the boiling point of water at atmospheric pressure. In one pattern of injector, where several overflows are connected, the author has seen the temperature of the water going to the boiler carrying 140 pounds of steam, reach 250° Fahr., with the overflow chamber closed only by a light check valve weighing but a few ounces.

As the proportion of water to steam is reduced, and the density of the jet diminished, it will be noticed that the value of the coefficient, as given in the last column, also grows less; when the water ratio was 1 to 13, as taken in the example at the opening of this chapter, this value was taken at 0.60, but when the weight of water per pound of steam is 7.02, this coefficient falls to 0.432.

The actual causes of the loss covered by this term cannot be definitely described; it is very probable that the whole mass of steam is not entirely condensed at the instant of impact, and there remains a certain amount of elasticity that produces a rebound and an interference with the motion of the particles of steam following. A jet of steam discharging into a pail of water will blow the water in all directions, and only a small portion of the steam will be condensed; the warmer the water, the greater this tendency, so that it is not at all surprising that with high temperatures of delivery, due either to warm feed water or insufficient feed supply, the coefficient of efficiency should have a lower value than when the conditions of working are more nearly normal. The oblique angle at which the water enters the combining tube is also disadvantageous, while the roughened surfaces of the tubes add their quota to the general sum of losses, which will be represented by the coefficient K.

It is exceedingly difficult, in fact almost impossible, to represent algebraically the conditions that obtain within the combining tube of an injector, and to frame an equation that will apply to all conditions or all types of instruments. The shape of the tubes, the conditions of the surfaces, and the proportions of the orifices, all introduce special considerations that would so complicate an equation as to invalidate its utility; fundamental relations can, however, be shown to exist, and simple equations given to prove the mechanical theory as already outlined; these will be followed by formulæ based upon the heat theory.

Taking the most elementary form of injector, one with a single set of tubes and but one overflow in the combining

THE ACTION OF THE INJECTOR.

tube, assume 1 pound of steam per second from the steam nozzle as the actuating force. This steam will have a velocity V due to the work performed in expanding from the initial pressure of the boiler to the low pressure in the combining tube. Its momentum will therefore be

$$\frac{V}{g} \quad \ldots \ldots \ldots \ldots \ldots \quad (17)$$

The feed water enters with a velocity v_2 due to the difference between its head h and the absolute pressure within the combining tube ; as these pressures must be reckoned above a perfect vacuum, the pressure in the tube becomes $(34-h_1)$, and if the water comes to the injector under a head h, the total head forcing it into the combining tube is $(34-h_1+h)$. Calling the weight of water W, the momentum of the entering water is

$$\frac{W v_2}{g} = \frac{W \sqrt{2g(34-h_1+h)}}{g} \quad \ldots \ldots \quad (18)$$

The sum of these two equations represent the momenta of the separate masses before they come in contact, but during impact and the condensation of the steam jet, there is a loss of momentum due to causes already outlined and indicated by the coefficient K. The sum of (17) and (18) is the momentum of the mass as it approaches the delivery tube ; its velocity must depend upon the difference between the absolute permissible back pressure—expressed in feet—and the pressure in the combining tube, or $(H-h_1)$. Therefore

$$v = \sqrt{2g(H-h_1)}$$

and the momentum of the combined mass is,

$$\frac{(W+1)}{g} \sqrt{2g(H-h_1)} \ldots \ldots \ldots \quad (19)$$

The complete equation of momentum is

$$\frac{K\left(V+W\sqrt{2g(34-h_1+h)}\right)}{g} = \frac{(W+1)}{g} \sqrt{2g(H-h_1)} \cdot (20)$$

Solving for W we obtain the ratio of the weight of water forced into the boiler per pound of steam,

$$W = \frac{KV - \sqrt{2g\,(H - h_1)}}{\sqrt{2g\,(H - h_1)} - K\sqrt{2g\,(34 - h_1 + h)}} \quad \ldots \ldots \ldots (21)$$

If h_1 is not known and cannot be readily found, it may be assumed to be about 24" vacuum or 7.3 feet absolute pressure when the delivery temperature is 150°, and 26", or 4.7 feet absolute, at 130°, although a slight variation will not materially affect the results. Upon h and h_1 depends the velocity of entrance of the water into the combining tube, and these terms do not appear in their true importance in the equation, as the area of entrance is not expressed, but is assumed to be of the most advantageous proportion; this is only the case in the self-adjusting or adjustable combining tube form of injector, shown in Figs. 1 and 2 or 3; as all the water must enter through this opening, whose area we will designate by B, the value of W in the ordinary fixed-nozzle pattern would be

$$W = 62.39\, B \sqrt{2g\,(34 - h_1 + h)} \quad \ldots \ldots \ldots (22)$$

If the supply is lifted, h must be negative; if the feed water flows, h is positive. To show the change in the value of W with different heights of lift ($-h$) the following tests of a fixed-nozzle injector, adjusted for its maximum capacity at 1 foot lift, are given in Table V. Steam was constant at 65 pounds pressure, and the area of the water entrance to the combining tube fixed; feed and steam valves wide open.

TABLE V.

Lift in Feet. $-h$	Capacity Cubic Feet. Q	Ratio Wt. Water to Steam. W	Lift in Feet. $-h$	Capacity Cubic Feet. Q	Ratio Wt. Water to Steam. W
1	108	14.54	8	87.8	11.83
2	107.2	14.43	10	81.8	11.02
3	104.9	14.13	12	76.9	9.92
4	101.9	13.73	14	66.9	8.97
5	97.9	13.18	16	60.8	8.13
6	94.8	12.76	18	58.3	7.72
7	90.4	12.43	20	51.6	6.89

In the self-adjusting injector, the area B changes automatically with the steam pressure, so that an increase in the

THE ACTION OF THE INJECTOR. 77

height of lift affects the capacity but slightly. A well known pattern of the double-jet type at 60 pounds steam, gave the following capacities at different lifts,

4 Feet Lift.	12 Feet Lift.	24 Feet Lift.
150.6 cubic feet.	134.6 cubic feet.	78.1 cubic feet.

In both (21) and (22) the head of water representing a perfect vacuum was taken as 34 feet; this corresponds closely to a temperature of 40°, but if the temperature of the feed rises, vapor of water is given off which reduces the vacuum correspondingly. The following table gives values that may be substituted for 34 in the above equation with different temperatures of feed water:

TABLE VI.

Feed Temp. t_1	Vacuum in Feet.	Feed Temp. t_1	Vacuum in Feet.
40	34	140	25.9
70	33	150	24.8
90	32.2	160	22.5
100	31.4	180	16.9
120	29.7	200	9.3
130	27.3	212	0.0

The chief criticism against the formula denoting the interchange of momenta, is that there is no term in the equation that indicates the use of condensible gases; as written, air or any other permanent gas could be substituted for steam and all the conditions satisfied; yet, as has been shown, the principle upon which the injector acts, precludes the use of all gases save those that may be reduced to a liquid form under the conditions that obtain within the combining tube of an injector.

The co-efficient K, which is such an important factor in the equation of momenta (21), seems to depend somewhat on t_2 but chiefly upon $(t_2 - t_1)$, the increase in the temperature of the feed water. As the heat of the water supplied rises, the weight of water delivered per pound of steam decreases, and the coefficient K has also a less value as is shown by Table VII; this table also gives the capacity, the value of W, and the density of the jet at 120 pounds steam, in a No. 8 Self-adjusting Sellers' injector, placed 1 foot above the feed level.

TABLE VII.

Feed Temp. t_1	Delivery Temp. t_2	Water to Steam. W	Cubic Feet per Hour. Q	Density of Jet. Δ	Coefficient. K
a) 65	144.5	13.53	278.7	0.8836	0.595
b) 80	160.0	13.26	274.2	0.8154	0.583
c) 115	202.5	11.63	239.2	0.7490	0.525
d) 130	226.0	10.36	213.0	0.6430	0.473

The falling off of the capacity with the increase in t_1, is caused by the reduced vacuum within the combining tube, and the change in the terminal velocity of the steam jet; this could be partially remedied by augmenting h in (22) as occurs in the double jet pattern of injector, but yet it may be stated as a general rule, that under conditions otherwise the same, the lower the temperature of the feed water supplied, the greater will be the capacity. In the case of the double jet type, the weight of water passing through a unit section of delivery tube is never as great as in a single jet injector, because the warming action of the first set, reduces the rate of condensation within the forcing combining tube, and lowers the density of jet.

It follows, therefore, that in order to give the same capacity as a single jet, the area of the delivery tube of a double-jet injector must be made considerably larger; the amount of increase required depends upon the proportions of the tubes, and may be anywhere from 13 to 80 per cent. The action of the first set, although disadvantageous as far as the warming of the water is concerned, yet produces a pressure at the entrance to the combining tube of the forcing set that increases its power of taking hot water; this pressure rises with the steam pressure and also with the temperature of the supply. In a series of tests made with an injector of this type, the following results were obtained,

Steam.	Pressure between Sets.	Steam.	Pressure between Sets.
40 lbs.	2 lbs.	90 lbs.	17 lbs.
70 "	8.5 "	120 "	27 "

when the feed temperature was 63 degrees, but when raised to 120 degrees the intermediate pressure was 22 lbs. at 90 lbs. steam, and 40 lbs. at 120 lbs. steam.

THE ACTION OF THE INJECTOR.

As the delivery of an injector depends directly upon the velocity of the jet and its density at the instant of passing the smallest diameter of the delivery tube, it follows that the capacity will be proportional to the square root of the terminal pressure plus a constant: if P is the boiler pressure as indicated by gauge, Q the capacity in cubic feet per hour, D the diameter of the delivery tube in millimetres, (1 millimetre = 0.0397″)—as many manufacturers use the French system—we have, approximately,

$$Q = 0.374\, D^2 \sqrt{P + 18} \quad \cdots \cdots \cdots \cdots (23)$$

This applies to self-adjusting injectors up to that pressure at which the maximum capacity commences to reduce, but does not refer to the double jet type, although a fair approximation for the other class.

In the following Table is given the weights of water that should be delivered per pound of steam by a well-designed injector at the steam pressures ordinarily used:

TABLE VIII.

Gauge Pressure.	W.	Gauge Pressure.	W.
10 lbs.	33 lbs.	100 lbs.	15.0 lbs.
20 "	29.2 "	120 "	13.6 "
40 "	23 "	150 "	12.6 "
70 "	17.8 "	200 "	10.3 "

Thus far the mechanical action alone has been discussed, but as the heat received by the feed water during the process of injection is one of the chief advantages of the system of feeding upon which the injector is based, the consideration of the thermal action will now be taken up.

In order to supply the energy necessary to give the final velocity to the particles of steam during expansion, an amount of heat must be absorbed corresponding to the work performed; according to the theory assumed under which expansion takes place, all the heat required must be furnished by the steam itself, and as the greater part of the heat contained in the steam is latent, a portion of the gas must be condensed; so that in its final state the steam is composed of a mixture of steam and water, in proportions which may be determined by equation (12). After expanding to the pressure within the combining tube, the steam strikes against the cold water, its momentum is checked and its energy of motion transformed to heat; this is at once absorbed by the feed, as is also the remaining sensible and latent heat in the steam mixture. It will be shown that all the energy that is apparently wasted, reappears in the form of heat, so that the injector may be considered as a theoretically perfect device for boiler feeding purposes, and its thermal efficiency as unity; the actual number of heat units that are lost by radiation that can be charged directly to the instrument itself is inappreciable compared with those contained in the delivered water.

Referring again to figures, take an injector which receives steam at 120 pounds and feed water at 68°, forcing 13 pounds of water per pound of steam against a permissible counter-pressure of 141 pounds, at a delivery temperature of 150°, and apply the analysis of the transfer of heat just outlined.

During the expansion of 1 pound of steam from 135 pounds to 4 pounds (absolute) there is an amount of energy developed that is represented by U in equation (11) or, by the term under the radical, in (13). Taking the latter equation, and substituting the figures previously used, and retaining the result in heat units,

$$x_1 r_1 + q_1 - x_2 r_2 - q_2 = 237.07 \text{ Thermal Units.} \quad \ldots \quad (24)$$

The water flowing into the combining tube possesses a small amount of energy which may be represented by

$$\frac{W}{778} \times \frac{v_2^2}{2g} = \frac{W}{778}(34 + h - h_1) = \frac{13}{778}(33 + 1 - 9.24)$$

$$= 0.75 \text{ Thermal Units} \quad \ldots \quad (25)$$

where 33 is the head of water in feet equal to the vacuum, h the head under which the feed flows to the injector and has a minus sign if the water is lifted; h_1 the absolute pressure in the combining tube expressed in feet; v_2 is the velocity of entrance; $h_1 = 1$ foot.

These two equations represent the motive power in the injector expressed in thermal units, one of which is taken as equal to 778 foot pounds of work. This energy is used for

THE ACTION OF THE INJECTOR. 81

forcing the mixture into the boiler with a velocity v, which requires

$$\frac{(1+W)}{778} \times \frac{v^2}{2g} = \frac{(1+W)}{778}(H-h_1)$$

$$= \frac{(1+13)}{778} \times (145-4) \, 2.355 = 5.97 \text{ Thermal Units} \ldots \ldots (26)$$

H is the absolute permissible counter-pressure in feet of head, and 2.355 is the head of water at a temperature of 150° equal to 1 pound per square inch from Table II. This value for the useful work performed—5.97 thermal units—is very small compared with the energy expended; adding (24) and (25) and subtracting (26)

$$(237.07 + 0.75) - 5.97 = 231.85 \text{ Thermal Units},$$

which must be absorbed by the feed water. The increase in temperature is therefore $231.85 \div 13 = 17.83°$ per pound of feed water due to the loss of actual energy of the steam at the time of impact. Expressed algebraically

$$\left[(x_1v_1 - x_2v_2 + q_1 - q_2) + \frac{W}{778}(34 + h - h_1) - \frac{(W+1)}{778}(H-h_1)\right]\frac{1}{W} . \quad (27)$$

which represents the increase in the temperature of the delivered water due to the absorption of the remaining energy of the steam jet.

When intimate contact is established between the steam and the water, condensation is effected, and the latent and sensible heat of the steam transferred to the feed. At this time, the heat in the expanded steam is

$$x_2 r_2 + q_2 + 32 - l_2 \ldots \ldots \ldots \ldots \ldots \ldots (28)$$

an equation of the same form as (24), but containing the final temperature of the delivery, as the temperature of the steam mixture cannot fall below that value. (The specific heat of water between 32° and 212° is so nearly constant, that the temperature of the feed is substituted for the "heat of the liquid," as the change simplifies the use of the formula, and the difference in the result is inappreciable.)

From (12) and also from the calculations made on page 64, we find the weight of steam remaining in 1 pound of initially
6

dry steam after expanding from 135 to 4 pounds absolute, is 0.8243; from the Steam Tables, r, the heat of vaporization $= 1007.2$; therefore,

$$\begin{array}{r} 1007.2 \times 0.8243 = 830.23 \\ 121.09 - (150° - 32°) = 3.09 \\ \hline 833.32 \text{ Thermal Units.} \end{array}$$

remaining in the steam, and given out to the 13 pounds of water during condensation; this will raise the temperature $833.32 \div 13 = 64.1°$, or in different form,

$$\frac{x_2 r_2 + q_2 - (t_2 - 32)}{W} = 64.1° \quad \ldots \ldots \ldots \ldots (29)$$

so that the sum of $64.1°$ and $17.83°$, will be equal to the total rise in the temperature of the feed as the water passes through the injector, or,

$$(t_2 - t_1) \ldots \ldots \ldots \ldots \ldots \ldots \ldots \ldots (30)$$

Adding these two values $17.83 + 64.1 = 81.93$ as the total increase in temperature, the delivery temperature will be

$$68.0 + 81.93 = 149.93°$$

which is the same as found experimentally.

Expressing these results algebraically, by adding (24) (25) and (29), subtracting the work required to force the mixture into the boiler (26), and equating to the increase in the temperature of the feed water (30),

$$\frac{1}{W}\left[(x_1 r_1 + q_1 - x_2 r_2 - q_2) + \frac{W}{778}(34 + h - h_1) - \frac{(1+W)(H-h_1)}{778} + (x_1 r_1 + q_2 - t_2 + 32°)\right] = (t_2 - t_1) \ldots (31)$$

Reducing and cancelling and changing the form in order to find the value of W, we have

$$W = \frac{x_1 r_1 + q_1 - t_2 - \dfrac{H - h_1}{778} + 32°}{t_2 - t_1 + \dfrac{H - 34 - h}{778}} \quad \ldots \ldots \ldots \ldots (32)$$

From this equation have disappeared the terms denoting the heat in the steam after expansion, and the important func-

THE ACTION OF THE INJECTOR.   83

tions remaining, are the total heat in the steam in its initial condition and the temperature of the feed and delivery. Let P denote gauge pressure; then, discarding unessential terms,

$$W = \frac{x_1 r_1 + q_1 + 32 - l_2 - (.003)\, P}{l_2 - l_1 + (.003)\, P} \quad \ldots \ldots (33)$$

This equation enables us to determine the weight of water delivered per pound of steam by any injector by simply observing the temperatures of the feed and delivery and the initial pressure of the steam, and substituting the proper values from any Steam Table; or, knowing the area of the steam nozzle, the discharge may be calculated by (12), which, multiplied by the ratio W, gives the capacity of the injector.

As brass is a good conductor of heat, there is always a certain amount of heat transmitted through the walls separating the steam and feed chambers, which increases the temperature of the delivered water. This heat, being abstracted from the steam, renders the discharge through the nozzle more or less wet, and reduces the quantity and velocity at the time of impact; if excessive, it seriously affects the ability of the injector to receive feed water at a high temperature, and also diminishes the range of capacities. This heat appears in the difference $(l_2 - l_1)$, and this term may be taken as a good test of the practical working qualities of an injector.

Probably the most satisfactory formula that could be devised to cover the action of the injector, would contain terms denoting the volumes of the steam, feed water and the combined jet, and based upon the exchange of momenta between the moving masses. If limited to the simple case of complete condensation of the actuating steam, this could easily be solved, and this analysis was in fact, followed in the discussion at the opening of this chapter, and the sizes of the various orifices calculated; but where only partial condensation is effected an uncertain element is introduced that can only be approximated by a formula based upon the phenomena supposed to take place within the combining tube and then introducing coefficients to make the results conform to actual experiments. The percentage of steam mixed with the water at the time of entrance to the delivery tube, and uncondensed, cannot easily be determined, as the temperature of the jet at that time is lower than when measured in the boiler pipe, and both the temperature and the pressure should be known for its correct determination.

The mechanical efficiency of the injector can be found by comparing the results of some of the equations already given; in the previous example used to illustrate the formulæ, the work of forcing 14 pounds of water against a head of 141 pounds, amounted to 5.97 thermal units, whereas the total energy expended, which is the sum of (24) and (25) = 237.82 thermal units, so that the efficiency is

$$\frac{5.97}{237.82} = 2\tfrac{5}{16} \text{ per cent.}$$

But the remainder of the energy is not wasted, but transformed to heat and absorbed by the feed water.

The maximum capacity of an injector is almost entirely a question of proper proportions and efficient action of the impinging jet, so that its value may be determined by equation (21). The *minimum* however is determined by the admissible temperature of the delivered water that will pass through the delivery tube without overflow; it should then be determined by (33) substituting for t_2 a value corresponding to the type of injector. For a single overflow injector, t_2 would be about 190°, but for two openings from the combining tube into the same chamber, probably 215°, at 60 or 80 pounds pressure. If, as an example, the case is taken of an injector receiving steam at 60 pounds, feed 70°, and a delivery temperature of 190°, we have from (33),

$$W_1 = \frac{x_1 r_1 + q_1 - (190° - 32°)}{190° - 70°} = \frac{898.8 + 276.9 - 158°}{120°} = 8.48$$

If the delivery temperature at the maximum capacity is 125°, the value of W would be 19.69, and the ratio of the minimum to the maximum would be

$$\frac{W_1}{W} \times 100 = \frac{8.48}{19.69} \times 100 = 43 \text{ per cent.,}$$

THE ACTION OF THE INJECTOR.   85

which is fairly good, although if the injector were so constructed that the highest delivery could reach 212°, the minimum would be reduced to 7.23, which would be 36 per cent. of the full capacity.

The highest admissible feed temperature is difficult to calculate without determining the special coefficients for each case, which can only be done experimentally. The warmer the feed water, the greater the weight necessary to condense a given weight of steam and, therefore, the heat equation must be used; on the other hand, the power of the steam for forcing water through the delivery tube is diminished on account of its reduced terminal velocity owing to the change in pressure within the combining tube, so that its momentum, and the rapidity of its absorption by the water jet is lessened and the density of the mixture reduced. Equating the equation of momentum (21) to (33) and solving for t_1 gives,

$$t_1 = \frac{t_2(KV-v) - (x_1 r_1 + q_1 + 32° - t_2)(v - Kv_2)}{KV - v} \quad \ldots \quad (34)$$

By way of illustration, take experiment (d) in Table VII; under these conditions, the mean vacuum in the combining tube is $10\frac{1}{2}''$, or $9\frac{1}{2}$ pounds absolute pressure, which gives a terminal velocity to the steam of 3050 feet per second. The water entering the combining tube has a velocity v_2 due to the low internal pressure which is equal to

$$\sqrt{2g(14.69 - 9.5) \times 2.31} = 27.76 \text{ feet.}$$

As we are determining the extreme capacity of the instrument, it will be assumed that the mixture has just sufficient power to enter the boiler; its velocity, v, will therefore be

$$\sqrt{2g(134.69 - 9.5) \times 2.42} = 139.7 \text{ feet;}$$

from the table, $K = 0.473$ and the delivery temperature $t_2 = 226°$; substituting these values in (34),

$$t_1 = \frac{226(0.473 \times 3050 - 139.7) - (1187.7 - 226 + 32) \times (139.7}{0.473 \times 3050 - 139.7}$$

$$\frac{- 0.473 \times 27.76)}{} = 129.5°$$

which agrees very closely with the result found.

The actual cause of the inefficiency of the injector from a mechanical point of view, is due to its dependence upon the principle of impact for the transfer of the actual energy of the steam to the moving water; this induces such an enormous waste of power that its use as a pump for raising water is out of the question except in such cases where economy of steam is no object, or where the heating of the delivered water may be advantageous. The momentum of a body is equal to the product of the mass by the velocity, and the resultant velocity after impact, is equal to the sum of the momenta of the two bodies divided by the sum of their masses; on the other hand, the actual energy is proportional to the square of the velocities. Even if there were no loss of momentum in the injector, yet there would be a dissipation of energy, owing to the reduction of velocity; if one pound of steam moving at a rate of 3000 feet per second strikes 9 pounds of water at rest, and imparts all its momentum, the resultant velocity of the mixture would be 300 feet per second, yet the energy of the jet of water and steam would only be $\frac{1}{10}$ of that of the initial steam; for,

$$\frac{\text{Final energy of jet}}{\text{Energy of steam}} = \frac{(300)^2(9+1)}{(3000)^2 \times 1} = \frac{1}{10}$$

so that the greater the difference between the velocity of the steam and the final velocity of the delivery, the greater the loss of actual energy; this also shows that the lower the terminal velocity, the greater may be the weight of water delivered, so that the equation simply measures the efficiency as a pump and not as a boiler feeder.

As an example, the efficiency of the jet will be calculated in the case of experiment (a) Table IV; the velocity of the steam is 3435 feet, of the entering water, 40 feet, and of the jet, 165.3 feet; the weight of water delivered per pound of steam, 12 pounds; therefore,

$$\left[\frac{(3435)^2}{2g} + \frac{(40)^2 \times 12}{2g}\right] K_1 = \frac{(165.3)^2}{2g} \times 13$$

$$K_1 = 3 \text{ per cent.},$$

and from this value there is not much variation throughout the ordinary range of steam pressures.

THE ACTION OF THE INJECTOR. 87

Probably the most satisfactory way to compare the injector as a boiler feeder with a direct acting steam pump, will be to take an example in which the actual coal consumption in the two cases can be compared. No allowance will be made for the advantages which may be claimed for either method in the way of convenience of operation or for economy of maintenance.

A steam boiler evaporates 10 pounds of water from and at 212° per pound of coal, and it is required to deliver 3000 pounds of steam per hour at 80 pounds pressure; water supply at temperature of 63°;

(*a*) *The Direct Acting Steam Pump.*

As the boiler receives the feed water at 63°, the actual evaporation is

$$\frac{966 \times 10}{1213 - 63°} = 8.40 \text{ lbs. water per pound of coal};$$

where 966 is the latent heat of one pound of water at 212°, and 1213 the total heat above 0° of 1 pound of steam at 80 pounds gauge pressure. The work of forcing 3000 pounds of feed water into the boiler per hour, or 50 pounds per minute, against a head of 80 × 2.31 feet is,

$$\frac{80 \times 2.31 \times 50}{33000} = 0.28 \text{ Horse Power},$$

where 2.31 is the height of a column of water at a temperature of 63°, equal to a pressure of 1 pound per sq. in. from Table II. According to experiments made by Prof. D. S. Jacobus, of the Stevens Institute of Technology, a direct-acting steam pump requires under these conditions 150 pounds of steam per horse power per hour; accepting this figure, an additional weight of feed water is necessary to supply the steam for the pump equal to

$$150 \times 0.28 = 42 \text{ lbs. per hour},$$

and to force this weight of feed water into the boiler requires 0.585 pounds of steam, determined as before, making an approximate total of 42.585 pounds of steam. This can be calculated precisely by the following equation, where y is the total weight of steam required for feeding the boiler:

$$\frac{\left(50 + \dfrac{y}{60}\right) 80 \times 2.31 \times 150}{33,000} = y = 42.596 \text{ lbs. feed per hour.}$$

The total weight of coal used by the boiler is

$$\frac{3000 + 42.596}{8.40} = 362.21 \text{ lbs. coal per hour.}$$

(b) *The Injector.*

Feed water is received by the injector at 63° and delivered to the boiler at 123.4°. The evaporation is, therefore,

$$\frac{966 \times 10}{1213 - 123.4} = 8.865 \text{ lbs. of water per pound of coal.}$$

The coal required to supply 3000 pounds of steam per hour,

$$\frac{3000}{8.865} = 338.41 \text{ lbs. per hour.}$$

Work of forcing 50 pounds of water per minute into the boiler,

$$\frac{80 \times 2.33 \times 50}{33,000} = 0.2824 \text{ horse power,}$$

where 2.33 is the height of an equivalent column of water at the new feed temperature of 123.4°.

As 1 horse power = 33,000 foot pounds, and 1 thermal unit = 778 foot pounds, 1 horse power is equal to 42.41 thermal units; therefore the coal required per hour for the work of feeding the boiler with the injector is

$$\frac{0.2824 \times 42.41 \times 60}{9660} = 0.74 \text{ lbs. per hour.}$$

Coal required to supply the steam that is applied by the injector to warming the feed water,

$$\frac{3000 \times 60.4}{9660} = 18.78 \text{ lbs. per hour.}$$

The total coal used in this case is

$$338.41 + 0.074 + 18.79 = 357.264 \text{ lbs. per hour.}$$

Saving of coal when the boiler is fed by the injector,

$$362.21 - 357.264 = 4.94 \text{ lbs. per hour.}$$

$1\tfrac{8}{10}$ per cent.

This saving of 4.94 pounds of coal per hour corresponds to the heat wasted in the exhaust steam from the pump and which is returned to the boiler by the injector.

CHAPTER VIII.

APPLICATIONS OF THE INJECTOR—FOREIGN AND AMERICAN
PRACTICE—DESCRIPTION OF VARIOUS PATTERNS
OF INJECTORS.

A TECHNICAL classification of the various patterns of injectors would be made according to the essential features embodied in the construction, similar to the method pursued in Chapter II, but as far as practical considerations are concerned, the distinctive feature is the presence or absence of a partial vacuum in the suction pipe at the time of starting. This may be produced by an auxiliary jet of steam, or by introducing the proper proportion between the area of the main steam nozzle and the upper overflow in the combining tube; either of these systems produces a vacuum in the feed pipe preliminary to starting, and constitutes the inherent characteristic of the *Lifting Injector;* the other class, requiring a pressure on the water supply, is termed *Non-Lifting*, or *N. L.* and permits a simplicity of construction that is seldom equalled in the other form; either class may be self-adjusting, single or double jet, although the N. L. class is usually made in the more elementary forms.

The injector as a boiler feeder has been successfully applied to all kinds of service, locomotive, stationary, marine, traction engines, etc., although it is to the first named that it has proved itself most useful, and for locomotive boilers has superceded all other devices. For stationary service, the non-lifting is used whenever a head of water is obtainable, principally on account of the simplicity of its operation. For large marine boilers, the injector is chiefly used in case of the disability of the pumps, but for small boats, the re-starting pattern has been extensively adopted. This kind is also used on traction, logging, and road engines, where its certainty of action and special adaptability, render it invaluable for the rough work to which such machines are applied.

The usual division of all the different forms of injectors resolves into a question of location relative to the water supply. A non lifting injector is usually tested under a head of about two feet, and will operate at all higher pressures by regulating the feed valve; this does not, however, give an increased capacity unless the steam pressure is raised. For the locomotive injector, the steam pressure under which it is tested and for which it is proportioned is about 120 pounds, while for stationary boilers, 60 pounds is substituted, although either arrangement will permit a wide range of working pressures. A lifting injector will of course operate with the feed water flowing to it, but the N. L. form will not work under the reverse condition unless it is first primed, and then only with reduced capacity, because the entrance area to the combining tube is too contracted.

Although there are many different styles specially designed for locomotive service, almost all correspond to one of two standards of capacity and position of the steam, feed and delivery connections; these are derived from the form of injector manufactured by two of the best known companies, viz., the Sellers' or P. R. R. standard, and the Monitor, or New York standard. This is advantageous in many ways, as it renders injectors interchangeable, and permits a careful and accurate comparison of the relative merits of the various kinds in use.

Neither in England nor on the continent is there any attempt at uniformity, nor is the construction limited to regular manufacturers; several railroads have designed and now make injectors for their own use, especially adapted to the requirements of their own motive power; of the two classes, the N. L. is more generally adopted, and it is usually of clumsy design and differs materally from those in use on American railroads. Fig. 18 is a drawing of the form used by the Great Eastern Railway of England and is applied to both passenger and freight engines. The com-

APPLICATIONS OF THE INJECTOR.

bining tube is heavily ribbed and forms the body, having its upper end spread into a double flange, to which are attached the steam and feed connections; on one side of the main casting is a lazy cock and on the other a heavy flange, the casing being made either right or left hand—depending upon the side of the locomotive for which it is intended—as it is bolted in a horizontal position to the frame, below the footboard. Large numbers of injectors answering to this general description are in use, yet no real standard prevails, and it appears that the English railway managers have not yet determined to their own satisfaction the most advantageous position upon the locomotive. Many manufacturers carry

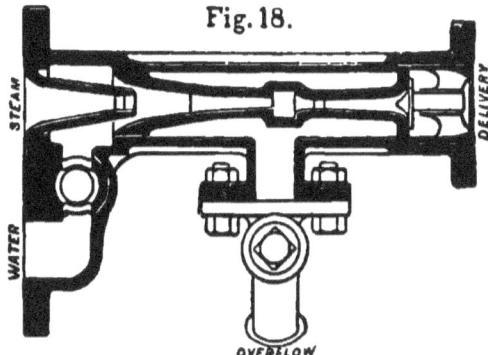

Fig. 18.

in stock all conceivable forms of casings for their standard tubes, adapted to be attached to any kind of locomotive in any possible position. Many of the original Giffard injectors are yet in use, and are still giving satisfaction; they are usually placed directly back of the fire box, and part way through the foot plate.

The most advanced English practice favors bolting the injector on the back plate of the boiler, with self-contained steam and check valves, and with steam and delivery pipes within the boiler. This arrangement is neat and compact, and as all parts are protected against accident, possesses some advantages over the practice in America of outside

92   THE GIFFARD INJECTOR.

connections. Objections urged against this system include the heating of the suction pipe and consequent difficulty in starting, and also the rapid closing of the delivery pipe within the boiler when the feed water contains lime or calcareous material in solution. This is due to the great difference in temperature between the water in the boiler and the water within the pipe, which causes a separation of the solids held in solution and a deposit on the interior sur-

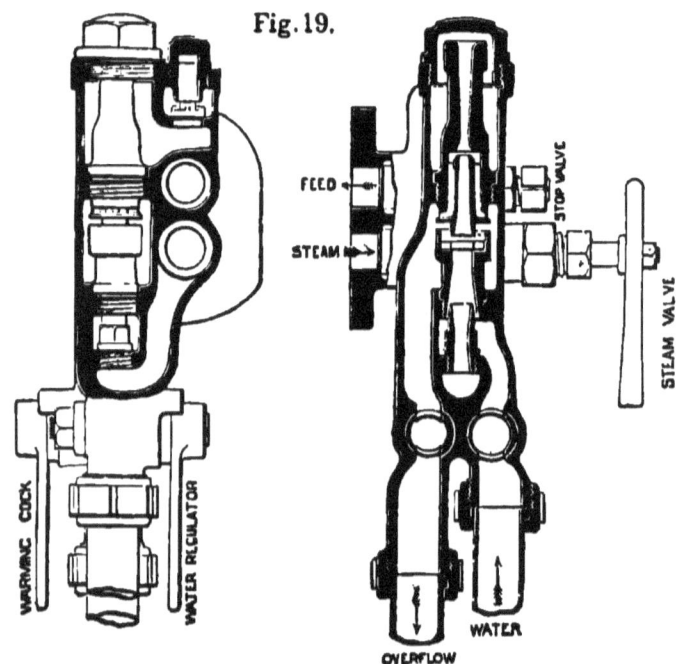

Fig. 19.

face of the pipe. With clean water these injectors have given good satifaction, and as they are growing in favor, one made by Gresham & Craven of Manchester is shown in Fig. 19. This form is inverted, the steam passing from the dome through the pipe A' to the chamber A; steam discharging upward from the nozzle a, raises the water without the necessity for a priming spindle, as the sliding combining tube b is forced up, exposing the large upper overflow of the

APPLICATIONS OF THE INJECTOR. 93

draft tube b'; the entrance of the feed water and the ensuing partial vacuum draws the tube down against its seat and prevents admission of air, while the jet passing through the delivery tube c is discharged at the front end of the boiler by the copper pipe C. This instrument belongs to the re-starting class and is easily primed; all the tubes may be removed when the boiler is under steam by closing the attached valves, and removing the top cap D.

Another form of injector that is somewhat similarly arranged, is extensively used on the London & Northwest Railway, and was designed by the Superintendent of Motive Power, Mr. F. R. Webb. It consists of an inverted set of tubes connected to a check valve on the back of the boiler, and lying as close to it as possible, but set low down and exposing but a short length of suction pipe to the heating action of the fire-box; the combining and delivery tubes are rigidly connected and moved to or from the steam nozzle by a convenient handle actuating a quick thread screw. The parts are set so that the water from the tank will flow through the overflow, so that no lifting device is necessary.

In France and in Italy, non-lifting injectors of the Friedman type are largely used—in the latter country, almost exclusively. From information furnished by the Compagnie de l' Est of France fully one-half of the 2441 injectors in use in 1887, upon its lines were of the Friedman pattern, and upon the Chemin de Fer de l' Etat, over 1000 of the same style were in service in sizes varying from 6 to 10 millimetres. During the last few years, however, many of the French railroads have been experimenting with injectors of American manufacture, and this has led to the extensive use by several of the companies of the Sellers' "Self-Acting" Injector, which is now replacing those of foreign design. This injector is shown in Fig. 26.

In Germany and Austria, a diversity of styles prevail, the greater number being non-lifting; even the Körting double jet is there applied beneath the foot board, while in this country it is used almost exclusively as a lifter. Other well known patterns are the Friedman, Hasswell, Pohlmeyer, Schaeffer & Budenberg, and Bohler.

Returning to American practice, it may be said without question, that the lifting injector is the standard class, as few railroads, except in some parts of the West and South, use the other forms. In certain localities where the feed water is impregnated with lime, non-lifters are retained on account of the facility with which they can be cleaned, and they are frequently used in combination with the other form, one being placed on each side of the engine. The chief disadvantage urged against this injector is that it is out of reach of the hand and eye of the engine-driver, and a sudden drop in the steam pressure or a careless opening of the lazy cock when the steam is shut off, may drain all the water out of the tank. This difficulty can be remedied by adding an extension to the overflow nozzle and raising it above the water level, and placing it where it may be continuously observed. This has been applied by the Nathan Manufacturing Co., of New York, to a modified form of their W. F. Injector, shown in Fig. 22. All non-lifting injectors should be placed so as to receive the feed water under as high a head as possible, as the action is thus rendered more certain; and the operation of starting, when the pipes are warmed by leaky steam valves, is much facilitated.

The almost invariable position of the lifting injector upon a locomotive is directly in front of the engine-driver, placed part way through the front cab frame so that the operating handles will be in convenient reach; the second injector usually occupies a similar position on the other side. On the Philadelphia & Reading R. R., where the camel-back engine is much used, the second injector is set through the back frame of the cab and within easy reach, so that both feeders are under instant command. Both the manner of placing the injectors upon locomotives, and the design of the instruments themselves, seem to be more carefully thought out in this country than elsewhere—unless the most recent English practice be an exception—and American manufacturers are still striving to bring the injector into

APPLICATIONS OF THE INJECTOR. 95

even more convenient shape and to render its action more efficient.

In order to show clearly the actual construction of the interior of the injector, selections have been made from the various patterns in general use and illustrations will be given of those most widely known.

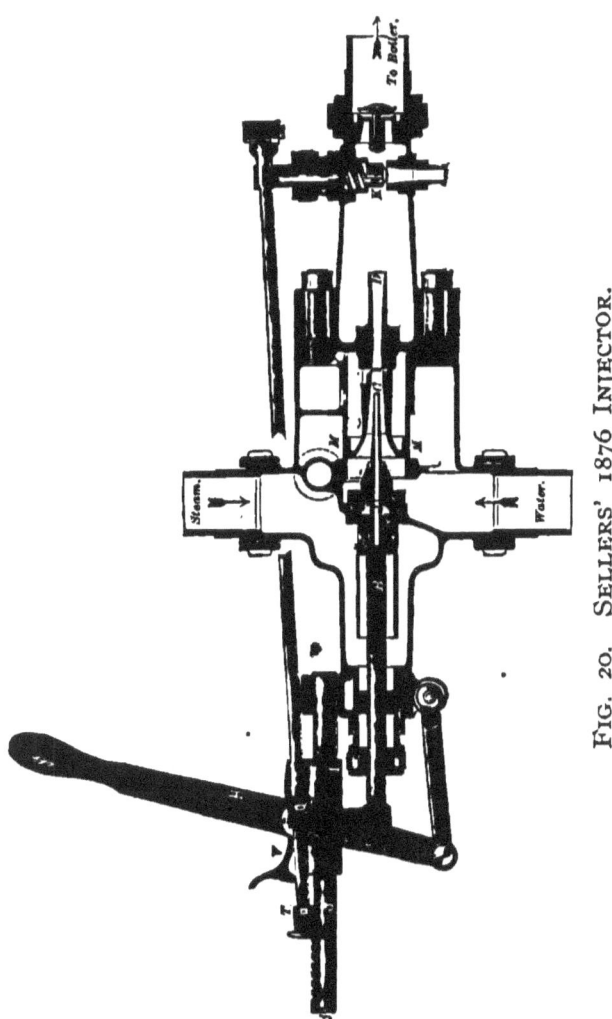

FIG. 20. SELLERS' 1876 INJECTOR.

THE GIFFARD INJECTOR.

Fig. 20 is a sectional view of the

SELLERS' INJECTOR OF 1876,

which contains the ingenious self-adjusting feature described and illustrated on page 15; it is conveniently constructed for operating and for repairing, and is started, regulated and stopped by means of a single lever, requiring no hand adjustment for any variation in the pressure of the steam, height of lift, or temperature of the feed water. The lifting is effected by drawing the lever H back a short distance until the resistance of the main valve X is felt; this admits steam to the hollow spindle C, which discharges through the delivery tube D, and produces a partial vacuum in the suction pipe; when water appears at the overflow, the lever is pulled all the way back, drawing with it the connecting rod L, closing the waste valve K and forcing the jet to enter the boiler; the regulation of the capacity is effected by adjusting the position of the lever by means of the steel latch V, on the guide rod J. Moving the lever changes the area of the steam nozzle by altering the distance which the taper spindle is inserted, and any variation in the weight of steam discharged, immediately induces an automatic movement of the combining tube which preserves the correct ratio between the water and the steam. An air chamber M, cored in the body, connects with the column of water in the suction pipe, so that all shocks and jars will be absorbed by the elasticity of the air, and will not be transmitted to the tube, and cause the injector to "break" or "fly off."

The following Table furnished by the makers gives the maximum capacity of this injector in cubic feet per hour:

Steam Pressure.	Size of Injector.							
	No. 3.	No. 4.	No. 5.	No. 6.	No. 7.	No. 8.	No. 9.	No. 10.
60 lbs. . . .	30	55	86	118	161	210	269	332
120 " . . .	40	70	112	160	218	288	357	445

APPLICATIONS OF THE INJECTOR. 97

THE MONITOR INJECTOR.

This is one of the best known of the locomotive injectors, and is a modification of Friedman's arrangement of tubes, with an improved casing. The combining tube is divided into two parts, forming an upper overflow at a point in the tube where the cross-section is about the same as that of the steam nozzle; this enables the injector to be started with great facility, and the air to be drawn from the suction pipe, by means of the independent ejector, quickly and promptly.

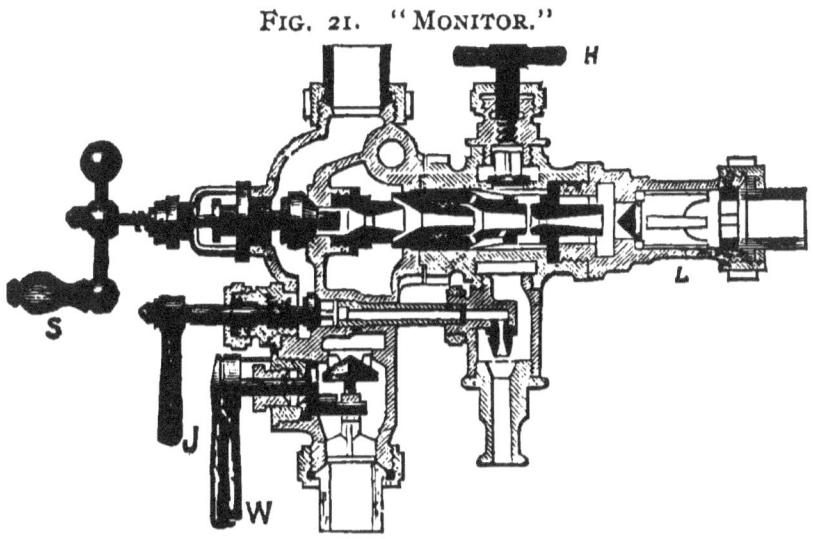

FIG. 21. "MONITOR."

The manipulation is as follows: opening the handle J, admits steam to the lifting jet which discharges through the waste pipe, creating a strong vacuum; when water appears at the overflow, the handle S opens the main steam jet and the water is forced through the delivery tube into the boiler; with high pressure steam, this draws all the water away from the lifting jet, and clear steam blows from the waste pipe which is checked by closing J; with low steam the water supply may have to be throttled by the regulating valve W in order to have the injector "run dry." This valve is also used to adjust the capacity to suit the requirements of the boiler, and permits of considerable variation.

7

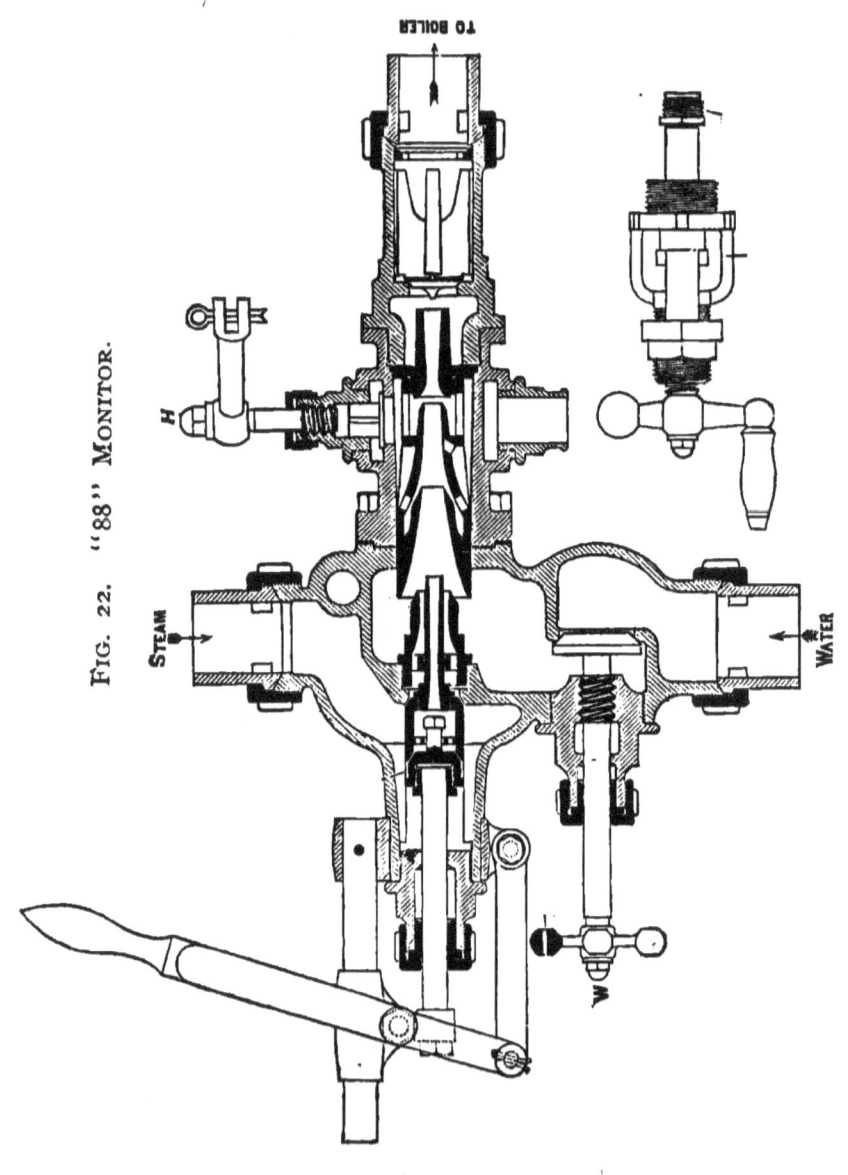

Fig. 22. "88" Monitor.

APPLICATIONS OF THE INJECTOR. 99

A newer form of this injector is that shown in Fig. 22, which can be operated with a lever or a screw handle on the steam spindle, and obviates the necessity for the starting jet *J*, by the use of the taper lifting-spindle inserted in the steam nozzle which discharges through the combining tube. The steam nozzle is made larger than in the form shown in Fig. 21, which enables it to operate at lower steam pressures, while a movement of the lever forces the taper spindle into the steam nozzle and reduces the area to a size better suited to high pressures; the range of steam pressures through which the injector will operate satisfactorily is much increased by this change.

The body is made in three parts, bolted and screwed together, and is heavily and strongly constructed throughout. The following table of capacities is reduced from that given in the catalogue of the Nathan Manufacturing Co., and applies to all the forms of the Monitor Injector, and also to the WF pattern, Fig. 23. This table gives the capacity in cubic feet per hour at a steam pressure of 140 pounds per square inch, feed temperature 60° to 70° Fahr., when the injector is placed at the usual height of lift on a locomotive:

Steam.	No. 4.	No. 5.	No. 6.	No. 7.	No. 8.	No. 9.	No. 10.	No. 11.	No. 12.
140 lbs.	73.3	126.6	166	240	306	386	453	506	560

Made by the same company is the

NATHAN WF INJECTOR,

designed to be placed below the foot-board with water flowing to it. The construction, as shown in Fig. 23, is very simple, and the steam nozzle and combining tube may be easily removed. Steam enters at the top of the casing and the water passes upward from the bottom through a cored passage to the entrance to the combining tube; the opening to the overflow valve is indicated in the cut by the circle between the combining and the delivery tube, and the

FIG. 23. NATHAN W F.

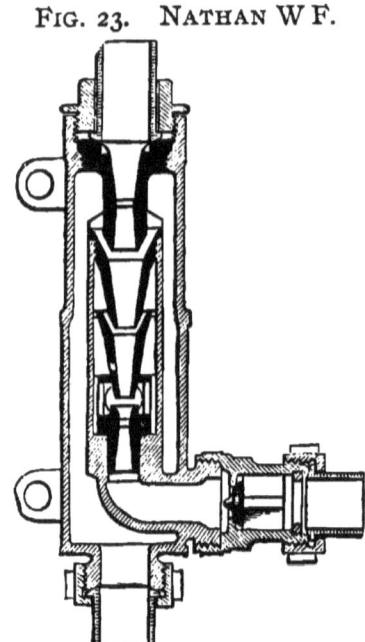

valve body may be placed on either side of the body, the opposite opening being closed with a screw cap. The proportions of the tubes and the arrangement of the overflows are similar to those of the Monitor pattern, and the list of capacities is the same.

A newer form of the non-lifting injector has been provided with a waste pipe extending above the water level of the tank and into the cab, where it may be under the constant observation of the engine driver, who is thus able to notice and check any tendency to waste. The tubes are arranged more in the form of those shown in Fig. 22, and a spindle is sometimes provided to facilitate starting.

Fig. 24 is a sectional view of the American form of the "Körting Universal," the

"SCHUTTE INJECTOR,"

which belongs to the double-jet type. It has no overflows in either the lifting or the forcing combining tubes, as the starting of the jet is effected by direct communication with the air through the compound waste valve placed vertically over the nozzle B. With this instrument the manipulation is as follows: Drawing the lever from A to B transmits the pull to the valve over the lifting steam nozzle by means of the bar held loosely between the caps K K; as the valve of the forcing set has a larger area, it is held by the steam pressure firmly to its seat, and acts as a fulcrum for the lever. After the appearance of the water at the waste pipe, the handle is drawn further out, and the vertical overflow valve, connected with the starting lever by means of a rod and bell-crank not shown, cuts off the outlet of the first set and diverts the water through the forcing combining tube. The second steam nozzle is then opened by the continued movement of the starting lever, which, during the latter part of its stroke, closes the waste valve; the jet then enters the boiler. Although the operation has been divided into several steps, the motion of the lever is practically continuous and the starting very simple. An elementary form of

FIG. 24. "SCHUTTE."

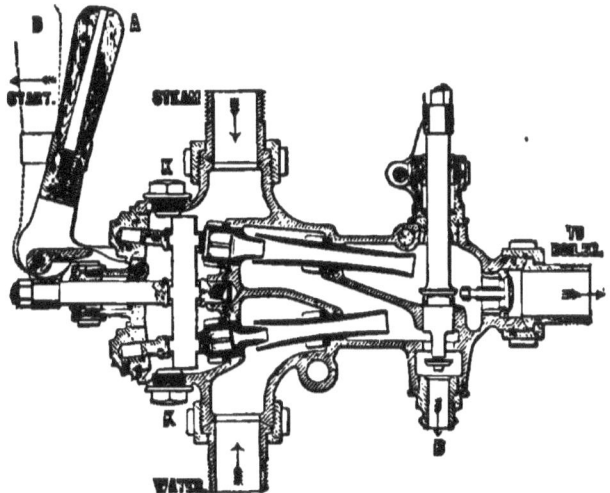

this injector was shown in Fig. 4, where the principle embodied in its construction was explained; reference has been already made to the peculiarities of construction which enable this type of instrument to receive water at a high feed temperature and operate through a wide range of steam pressures; at 60 pounds steam, the permissible temperature of the feed water, lifted 5 feet, is 150°, and at 125 pounds steam, 135°.

The capacity in cubic feet per hour is shown by the following table, based upon the following conditions: Feed, 80°, lift 4 feet; at 65°, the values will be increased about 5 per cent.; on the other hand, when the height of lift is 10 feet, there will be a reduction of the capacity of 10 per cent.; at 15 feet, 25 per cent.

Steam.	No. 1.	No. 2.	No. 3.	No. 3½	No. 4.	No. 5.	No. 6.	No. 7.	No. 8.	No. 9.	No. 10.
30	13	20	33	46	60	75	90	128	165	195	240
60	16.5	25	39	57	76	98	120	150	195	240	285
120	21	29	48	66	86	111	136	195	255	300	345
150	24	31	53	73	96	123	150	215	282	330	380

In the injector illustrated, the capacity is adjusted to the requirements of the boiler by means of a valve placed in the suction pipe, but the more recent improvements include a small taper spindle inserted in the steam nozzle of the lifting set, and adjusted by a handle directly under the starting lever. This varies the flow of steam in the lifting set, changing the amount of water delivered, and materially reduces the steam consumption at the minimum capacity. It is so arranged, however, that the lifting power of the injector is in no way affected, for the water can be raised and forced into the boiler even when the regulating lever is set for the lowest delivery of the instrument.

THE BELFIELD INJECTOR

also belongs to the double-jet type, but differs from the previous form in having all the tubes in the same axial line; suction is produced by the hollow cylindrical spindle, which is, however, never entirely withdrawn from the first steam nozzle, so that the entering water receives impulses from both the central and the surrounding jet. After passing the first delivery tube, the feed water meets the discharge from the annular second jet, opened by a second movement of the starting lever; the water then enters the boiler, but requires, in order to obtain more positive action and a fair minimum capacity, the tight closing of the waste valve, effected by the final motion of the lever. Capacities at 125 pounds and 5 feet lift are given in the following table; the size number is arbitrary, and is based upon the capacity. The injector will

APPLICATIONS OF THE INJECTOR. 103

operate over a wide range of steam pressures, and gives a fair minimum.

Steam.	No. 3.	No. 4.	No. 5.	No. 6.	No. 7.	No. 8.	No. 9.	No. 10.
125 lbs. . . .	40	73	113	160	220	290	366	452

FIG. 25. "BELFIELD INJECTOR."

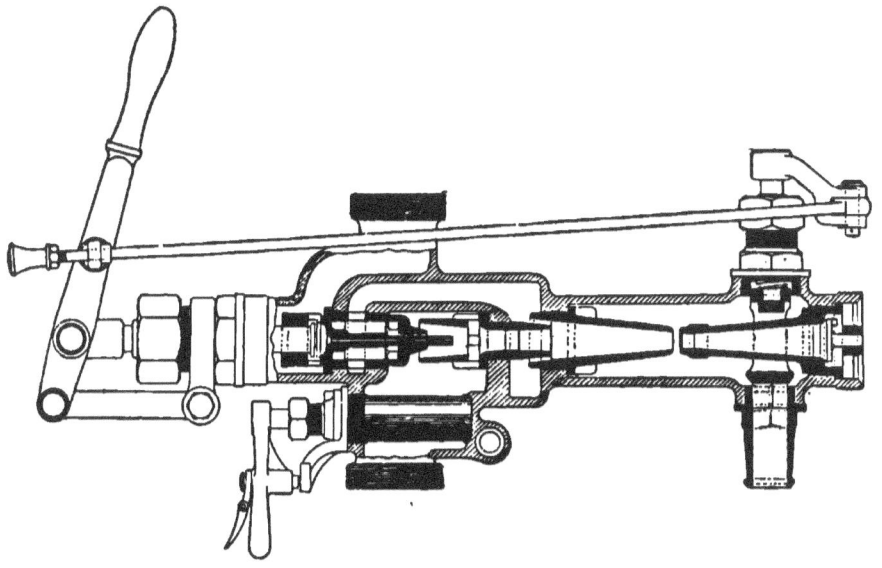

A notable characteristic of the

SELLERS' SELF-ACTING INJECTOR OF 1887

is the simplicity of its construction; its special features are obtained by the arrangement of its overflows and the proportions of its tubes, instead of by moving parts or special valves. In common with the self-adjusting injectors already described, and double-jet injectors in general, it adjusts and maintains the proper supply of feed water for each pressure of steam, and, in addition, re-starts automatically after a temporary interruption of the water or steam supply.

As may be seen from Fig. 26, these results are obtained by means of the following arrangement of parts: A small

104   THE GIFFARD INJECTOR.

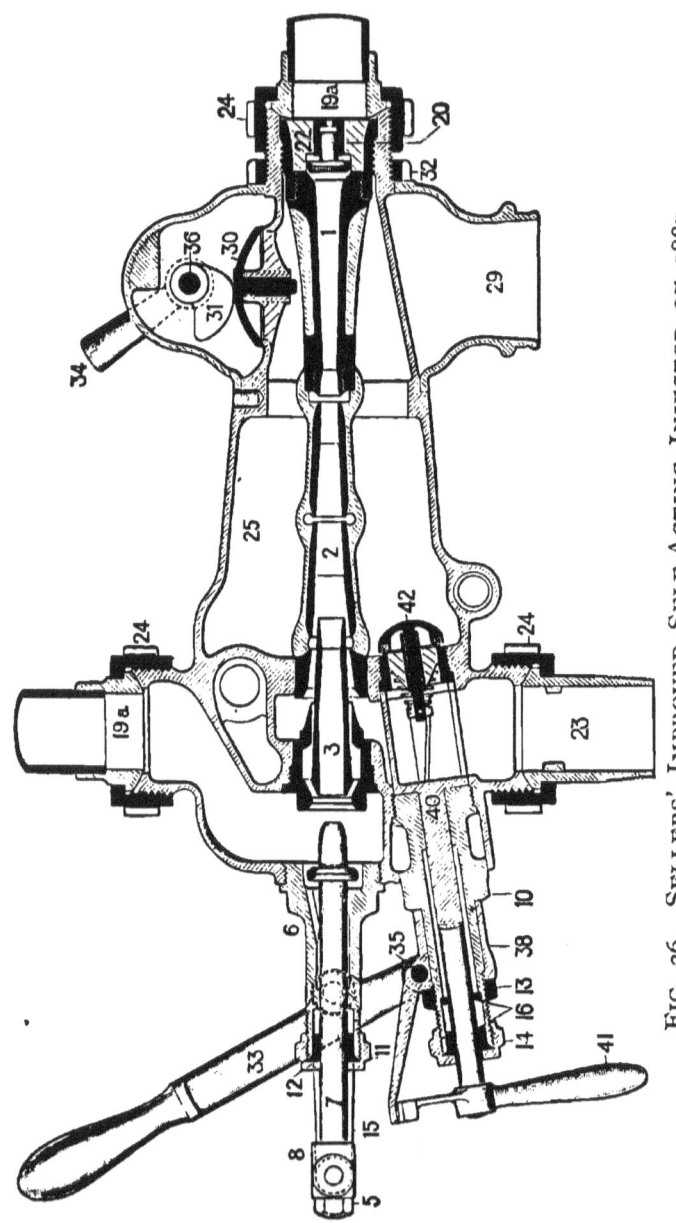

Fig. 26. Sellers' Improved Self-Acting Injector of 1887.

annular steam nozzle discharges through the annular area surrounding a central forcing steam nozzle, and through the overflow spaces in the combining tube ; the proportions of these parts are such that the lifting steam nozzle can always produce a suction in the feed pipe even when there is a discharge from the main steam nozzle, and it is this fact that establishes the re-starting feature. When the feed water rises to the tubes, it meets the steam issuing from the lifting nozzle, and is forced by it in a thin sheet and with a high velocity into the combining tube of the forcing set, without the necessity for the interposition of a divergent delivery tube. The water there comes in contact with the main steam jet, receives an additional impulse, and the mixture, reducing in cross-section as the velocity is accelerated, rushes past the overflow openings in the combining tube and passes through the delivery tube into the boiler.

The self-adjusting action is accomplished by the change of the capacity of the first set with the steam pressure, assisted by the influence of the partial vacuum in the combining tube, which, being communicated to the surrounding overflow chamber through the apertures in the tubes, assists in drawing an additional supply of feed water from the suction pipe whenever the requirements of the jet are increased by a rise of the boiler pressure. This maintains a constantly increasing capacity with the pressure, and gives an exceptionally large range of boiler pressures with which the injector will work efficiently.

On account of the low temperature of the delivered water, the weight of water forced into the boiler per pound of steam is unusually high, especially at locomotive boiler pressures ; this small consumption of steam permits the grading of the capacity through a wide range, and at 120 pounds steam, when lifting 1 foot, the capacity may be reduced by throttling the feed valve to 36 per cent. of the maximum under the same conditions ; this range of available capacities enables the injector to be used to feed the boiler continuously by simply adjusting the water valve to suit the requirements of the boiler.

Not only is the range of capacities dependent upon the minimum expenditure of steam to perform the work required, but also the highest admissible temperature of the feed water; further, as this is dependent also upon the limiting temperature of the mixture that will pass through the narrowest part of the delivery tube without overflow, it follows that an injector that can maintain a continuous jet at a temperature of 250° at the minimum capacity without wasting from the overflow nozzle, freely opening to the air, is able to use very hot water. Actual tests when lifting 1 foot at 120 pounds steam, show a limiting temperature of 124° for the automatic action, and 136° when the waste valve is held to its seat by throwing over the cam lever; these results were obtained with fairly dry steam and without special care in handling; regulation of the feed valve is the only movement required to obtain the minimum capacity.

The most recent improvement in this form of injector is shown in Fig. 26. A new check valve has been introduced between the water supply and the overflow chamber for the purpose of admitting a supplemental supply of water whenever there is a partial vacuum therein, caused by the incomplete condensation of the steam in the upper part of the combining tube. It was found that an insufficient amount of water passed through the annular lifting tube at pressures above 160 lbs., and produced a high vacuum within the overflow chamber. This vacuum is now utilized to draw in an additional supply of water, which enters the lower tube and passes with the jet into the boiler, increasing the capacity about 24 per cent. at 180 pounds pressure. When the vacuum in this chamber is less than that in the suction pipe this check valve closes automatically, preventing any steam or hot water entering the suction pipe during the operation of priming.

That it is very efficient at the high pressure is shown by the accompanying table of capacities, which gives rapidly increasing figures as the pressure is raised. The range is good: 57 per cent. at 200 lbs., and 63 per cent. at 120 lbs.

APPLICATION OF THE INJECTOR. 107

A full and complete test is given, page 134. This injector is specially designed for locomotive service, and occupies but little room in the cab; upon removing the overflow sleeve after unscrewing the jam nut, the body may be placed through a round hole in the front cab frame, allowing only the steam and feed pipes and manipulating handles to project into the cab. Each size of injector is named from the diameter of the delivery tube expressed in millimetres, and the capacity is given in the following table in cubic feet per hour at the ordinary steam pressures employed; height of lift. 5 feet.

SIZE.	30	60	120	160	200
5 4/10	67	89	121	133	138
6¼	98	129	176	193	202
7½	130	172	234	258	268
8½	167	221	301	332	345
9½	209	276	376	410	425
10¼	255	338	460	505	527
11½	305	405	551	607	630

THE METROPOLITAN INJECTOR.

This injector is of the double-tube class. It is provided with a *closed overflow*, and designed with a view of obtaining a double-tube injector without outside attachments, rendering the construction simple. The lifting apparatus lifts the water, delivers it under pressure to the forcing apparatus, which in turn discharges it into the boiler. There is a double steam valve for successively admitting steam to the lifting apparatus and to the forcing apparatus; to the stem of the main steam valve the overflow valve stem is rigidly connected by means of a cross-head. The overflow valve stem passes

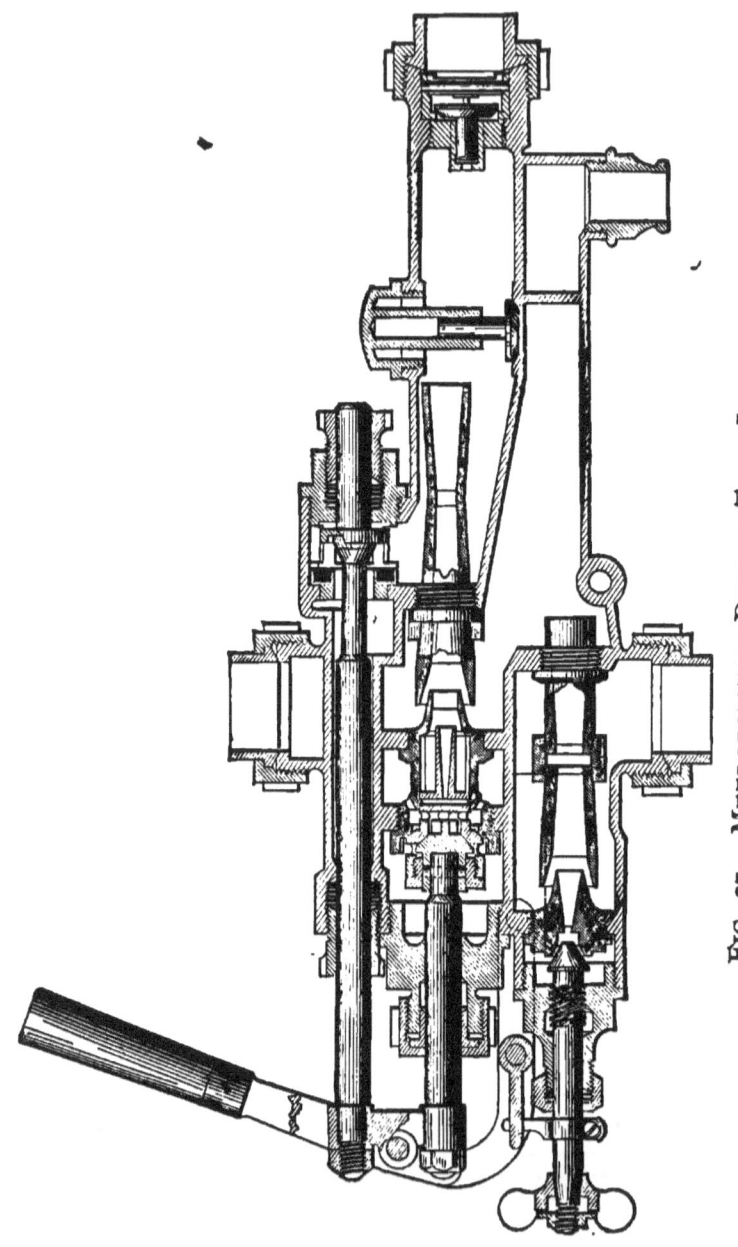

Fig. 27. Metropolitan Double Tube Injector.

APPLICATION OF THE INJECTOR. 109

through a cored passage in the injector, which opens directly into the atmosphere through the overflow passage. By the form of valve-stem used, and the means for operating it, an overflow valve of the simplest type can be used.

The overflow valve seat is made removable from the casing, thus facilitating repairs. The manufacturers state that this injector can be started with 25 lbs. steam pressure, and without any adjustment or regulation of any kind, can be operated at steam pressures up to 275 lbs. Also, that these injectors will take feed-water at 145 to 150 degrees F., with 100 lbs. steam pressure; at 142 degrees F. with 125 lbs. steam pressure; at 135 degrees F. with 150 lbs. steam pressure; at 130 degrees F. with 175 lbs. steam pressure, and at 125 degrees F. with 200 lbs. steam pressure. The capacity is regulated by increasing or decreasing the amount of steam to the lifting apparatus, by adjusting the small hand wheel below the operating lever

Its operation, briefly, is as follows: The lever being drawn back slightly opens the steam valve to admit steam to the lifting apparatus; steam passing through the lifting apparatus raises the automatic valve, which gives instant relief, and thus enables the injector to lift the water promptly. As soon as the water is lifted it takes the same course as the steam, passes out through the overflow, thus filling the water passages of the injector. As soon as water appears at the overflow the lever is drawn back admitting steam to the forcing apparatus, which seats the automatic overflow valve, owing to the increased pressure on top of this valve, and finally closes the overflow valve of the forcing apparatus, thus turning the water from the overflow into the boiler.

The injectors already illustrated comprise the most prominent in use upon American locomotives, and contain special features, which are in many cases duplicated, though perhaps in a less efficient degree, in other patterns; special mention, however, should be made of the Hancock Inspirator, Garfield, Mack, Eynon Evans and Ohio Injectors, but it would

110 THE GIFFARD INJECTOR.

occupy too much space to illustrate these and all the other patterns in use, and those given embody the most successful application of the essential principles. Many of these injectors are also used upon stationary boilers, yet they are more generally applied to railroad service.

The following have been selected from the great variety of injectors designed for stationary, traction and marine boilers as containing special features of interest; almost all are well known, and are carefully designed and accurately manufactured.

FIG. 28.

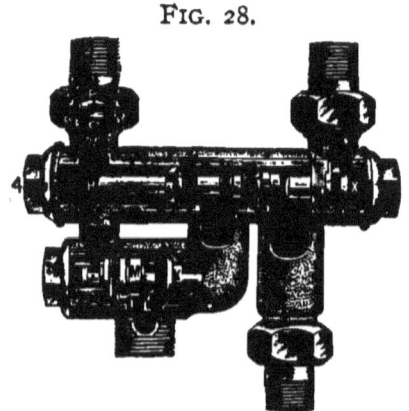

Almost all single-jet injectors designed for stationary or marine service, are arranged to start through a lower overflow or aperture placed between the mouth of the delivery tube and the lower end of the combining tube. An exception, however, is the

PENBERTHY SPECIAL

which is shown in Fig. 28, and described by the manufacturers as an Auto-positive Injector. The usual upper overflow, characteristic of re-starting injectors is used, but the combining and delivery tubes are made continuous, without spill holes or separating space. The jet, therefore, starts through the delivery tube, as the final or pressure chamber H has free access to the air when the valve L is raised from its seat on the bushing M whenever the valve K opens. The

APPLICATION OF THE INJECTOR. 111

action of starting is as follows: Steam is admitted through the upper right-hand branch to the nozzle X and discharging through the draft-tube G, produces a vacuum in the suction pipe; the discharging steam forces the automatic valve K against the end of the pressure valve L, which is thus held open to permit free outlet of steam and water from the chamber H; this is made possible by having the area of K about twice that of L. When the correct proportions of water and steam are admitted and the jet is formed, the pressure in the chamber containing the upper overflow is changed to a partial vacuum, which draws the automatic valve K to its seat and allows the pressure valve L to close, compelling the feed to pass under the check valve P and into the boiler.

The construction of the injector is very simple and easily understood. As it has no lower overflow, the action is similar to that of a positive or closed overflow injector, so that the overflowing temperature corresponds to the breaking temperature of the ordinary single jet pattern. The range of steam pressure is said to be good; the minimum capacity is obtained by throttling the water supply, for which purpose a valve must be placed on the suction pipe.

A sectional view of the

SELLERS' RESTARTING INJECTOR.

is given in Fig. 29. The branches for steam, water supply and delivery to the boiler are conveniently arranged, so that all the pipes may be placed close against the boiler wall. The overflow is directly under the water branch and can be provided with a drip funnel and discharge pipe, without bending or springing the other pipe connections. The steam nozzle and delivery tubes are *screwed* into the body, and do not depend upon the pressure of the steam or of the delivery to hold them in place, so that there is no danger of leakage at these important shoulders. The body and tubes are constructed of the best bronze and are designed to give the longest service with the least amount of attention and repair.

The injector is simply constructed, and contains but few

parts. It is perfectly automatic in its action, restarting instantly after a temporary interruption of the steam or water supply. It raises the feed promptly on long lifts, with hot or cold pipes, and gives a good range of capacities. Steam enters at the top and passing down through the steam nozzle, No. 3, discharges through the draft tube into the overflow chamber and thence to the air, lifting the water to the Injector. The partial vacuum caused by the condensation of the steam within the combining tube raises bushing No. 5 up

FIG. 29.

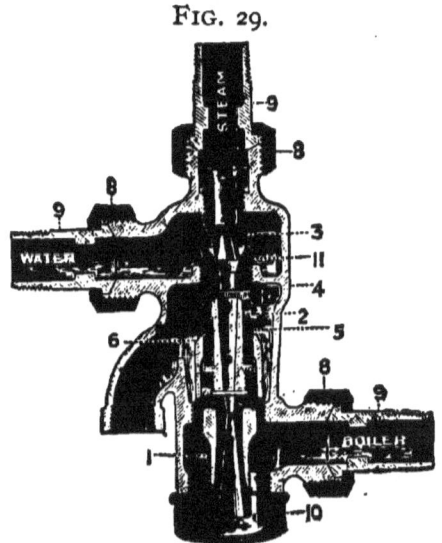

against the draft tube and holds the lower bushing, No. 6, against the delivery tube, thus preventing the admission of air.

Upon removing the cap at the lower end of the body the end of the delivery tube will be seen projecting *below* the lower face of the body, so that a monkey wrench may be used to unscrew this tube, drawing out the tubes and the overflow bushings at the same time.

The size numbers of these injectors are based upon the diameter of the delivery tube expressed in tenths of millimetres; No. 16, for instance, is $1\frac{6}{10}$ millimetres in diameter.

APPLICATION OF THE INJECTOR.

For all ordinary cases, this injector requires no adjustment and will work satisfactorily from 40 to 100 pounds steam, water flowing from a tank or from the city mains, or when lifting any distance up to 15 feet. If it is necessary to use the injector continuously at 15 or 20 pounds, the small ring placed under the steam nozzle should be removed, and the steam nozzle screwed down closer to the draft tube. For 150 pounds steam and very high lifts and feed temperatures, a greater distance is required between the steam nozzle and the draft tube, and a supplemental ring, supplied with the injector, set in place, permitting a wide range of working pressures. The range of capacities is good, and warm water can be fed to the boiler.

Fig. 30 is a section of the

"RUE INJECTOR" (LITTLE GIANT)

with which the experiments were made which were tabulated on page 47. This instrument has a movable combining tube by which the water area may be regulated for any steam pressure or desired change in the capacity, and this is

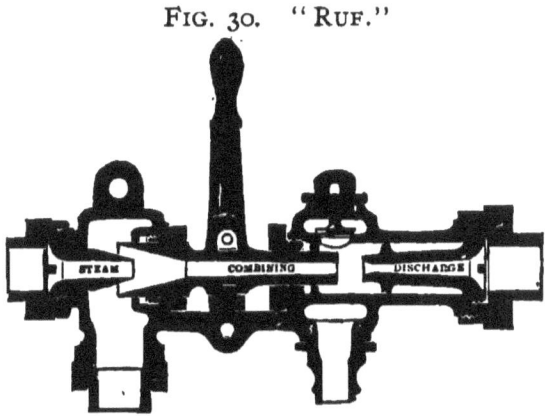

FIG. 30. "RUF."

effected by moving to the right or left the hand-lever fulcrumed on the body. When arranged as a lifting injector, a separate steam jet is added, either in the form of a central spindle discharging through the combining tube or a special nozzle placed in the waste pipe. The advantage of the adjustable water area has already been described in detail, so that the merit of this feature does not need further remark. Outside stuffing boxes permit absolutely tight joints between the water and overflow chambers and the atmosphere, so that no leakage of either steam or water need occur.

Another simple form of injector, but which has fixed nozzles, shown in section in Fig. 31, is the

SELLERS' F. N. INJECTOR OF 1885.

In this injector the upper and lower overflows are so arranged in the combining tube that it is automatic in its starting action, and gives, in addition, a strong vacuum, suitable for high lifts, when the steam displacing spindle is run slowly out. The form of this spindle also allows of a regulation of the steam supply when the water valve is throttled, lowering the delivery temperature so as to give at

FIG. 31. SELLERS' F. N. INJECTOR OF 1885.

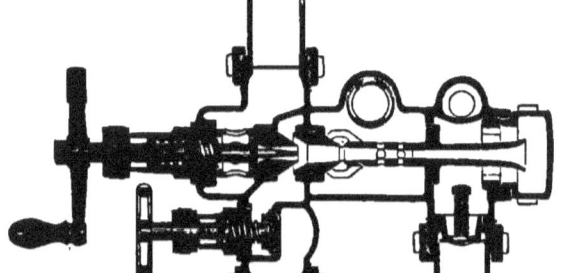

60 pounds steam, a minimum capacity of 25 per cent. of the maximum. The list of capacities conforms closely to that of the Sellers' 1876 pattern, yet the permissible temperature of the feed water is higher at the lower steam pressures; at 65 pounds, for instance, feed at 136° may be used when flowing to the injector under a small head, without interfering with the automatic action, and this may be increased to 150° if the waste valve is screwed down. The

APPLICATIONS OF THE INJECTOR.

tubes are removed from the body for cleaning or repairs by the motion of unscrewing the end cap, without the necessity of breaking any of the pipe connections; water, steam and delivery check valves are all contained in the body, so that the injector can be used either as a lifter or a non-lifter.

An ingenious device by which the area of the water entrance to the combining tube can be varied without the use of a stuffing-box, is illustrated in Fig. 32, a sectional view of the

METROPOLITAN INJECTOR.

The operation is as follows: Rotating the handle K opens the steam valve and lifts the feed water by the strong suction produced by the discharge around the plug M passing through the draft tube; when this plug is entirely with-

FIG. 32. "METROPOLITAN."

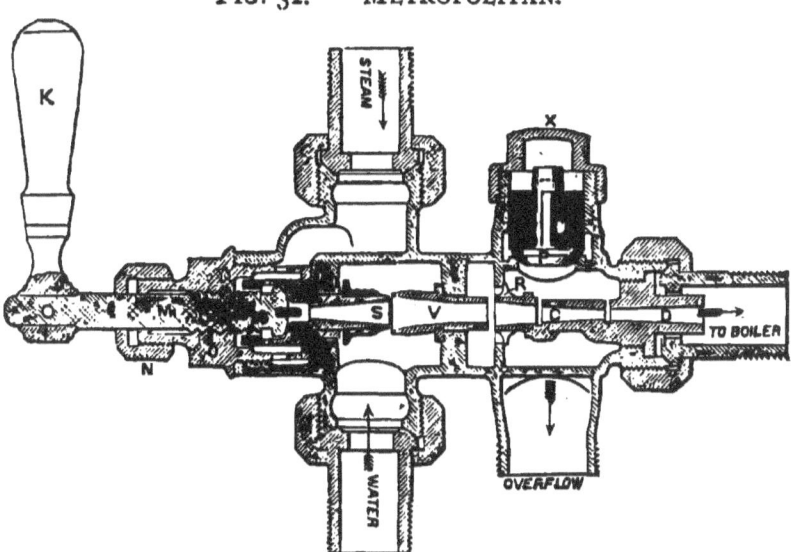

drawn, the full area of the nozzle is given and the jet is driven into the boiler. If the steam pressure is high or the height of lift is great, the usual supply area of the combining tube will be insufficient, and the spindle must be drawn against the rear stop, pulling the steam nozzle with it until the collar on *S* strikes against the back wall of the water chamber. For low steam, the spindle is screwed until the collar of the steam nozzle reaches the inner stop, when the pressure of the steam holds it fifmly in its place. Care must be taken to preserve tight joints between these faces, so that no leakage can occur between the steam and feed chambers, as this will not only reduce the lifting powers, but also the capacity of the injector.

An English invention that has been used in this country for some years on portable, marine and stationary boilers, is the

GRESHAM RE-STARTING;

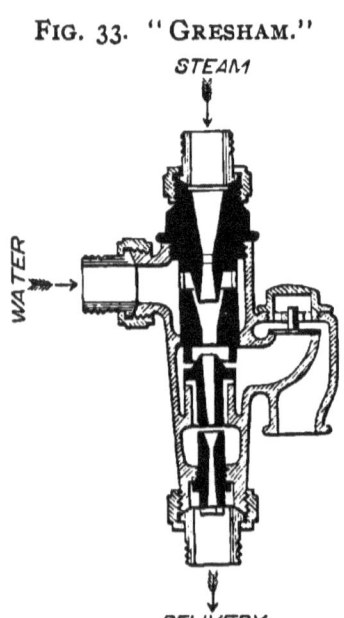

FIG. 33. "GRESHAM."
STEAM
WATER
DELIVERY

its interior construction is similar to the locomotive injector shown in Fig. 19. It has the same form of movable combining tube which closes the upper overflow when the injector is in action, and therefore requires no priming device, as the suction produced by the discharge of the steam through the draft tube is exceedingly effective.

An ingenious invention for permitting a free discharge of steam through the combining tube, yet giving an absolutely continuous tube to the moving water jet, is the flap nozzle system, invented and patented by Hamer, Metcalf & Davies, of Manchester, England, in 1880, and shown in Fig. 34 as a feature of the

MANHATTAN INJECTOR.

This form of tube is more extensively used in the injectors using exhaust steam than the usual form of overflow openings. The tube is cut longitudinally at its middle section up

APPLICATIONS OF THE INJECTOR.

to a point near the steam nozzle, where the cross section of the tube is sufficient to permit easy exit for water or steam. The force of the steam striking against the loose hinged portion, raises it, exposing a large area of discharge; as soon as the water enters the tube, the partial vacuum caused by the condensation of the steam draws the tube shut and holds it in that position as long as the jet is continuous. This system is specially advantageous when applied to exhaust injectors, where free discharge is a pre-requisite condition, as there should be no possible danger of the feed water

FIG. 34. "MANHATTAN."

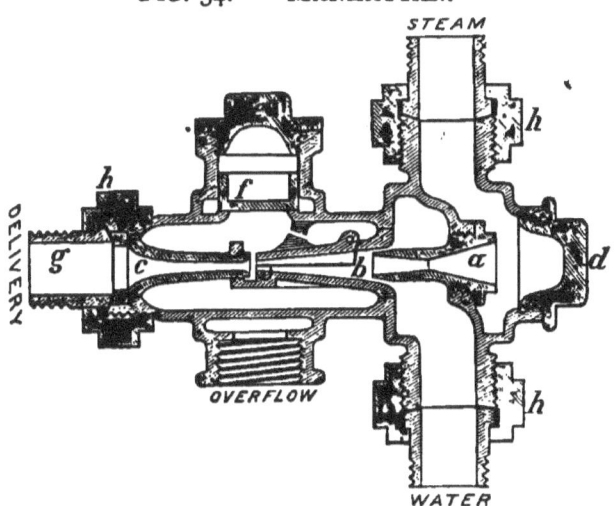

rising in the steam nozzle when the jet breaks, as it might work its way up into the cylinder. The illustration shows a live steam injector, re-starting on short lifts, but intended primarily as a non-lifter; a steam valve and displacing steam plug can be inserted in place of the cap d if desirable. All the tubes are easily removable, and care should be taken to allow no scale or sediment to be deposited upon the surfaces of the closing jaws, as this is apt to interfere with the tightness of the joint and the correct alignment of the tubes. The different parts are clearly shown in the cut—a, b, c, being the steam nozzle, combining tube and the delivery tube; f is the overflow valve to prevent air being drawn to the boiler by the suction of the jet at the lower overflow or through any leak between the joints of the tubes.

The same form of combining tube is applied to the Schaeffer & Budenberg

EXHAUST STEAM INJECTOR,

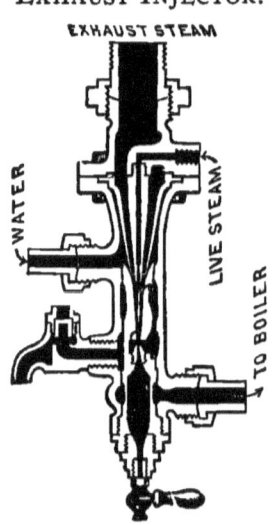

FIG. 35.
EXHAUST INJECTOR.

illustrated in Fig. 35. Here the steam nozzle is enlarged to meet the new conditions, and a central spindle is added, which, when the pressure of the boiler into which the feed water is delivered exceeds 75 pounds, is made hollow, so as to admit a supplemental jet of live steam from the boiler. The feed water should flow to the instrument and its temperature be as low as possible, never, under any conditions, above 90°; with the supplemental jet and cold feed water, the injector will feed against pressures up to 150 pounds, and without it, against 90 pounds; increasing the temperature of the feed diminishes these pressures somewhat.

THE PENBERTHY INJECTOR,

which is shown in Fig. 36, also belongs to the re-starting class, as it is arranged with an overflow near the steam nozzle, which is closed when the injector is in action by the loose washer T sliding upon the outside of the combining tube Y. Steam discharges from the nozzle R, flows through the larger opening at the mouth of the suction tube S, and raises the water to the feed chamber, the inlet branch being at the side of the body facing the observer, and therefore not appearing in the drawing; the partial vacuum caused by the condensation of the steam raises the washer T to its seat in the same manner as the combining tube in the Gresham Injector, so that the two overflows are separated and no air

APPLICATIONS OF THE INJECTOR. 119

can enter, even if the waste valve P should leak; this valve is made large enough to permit free discharge for the steam without producing any back pressure in the overflow chamber, as this would interfere with the lifting power of the

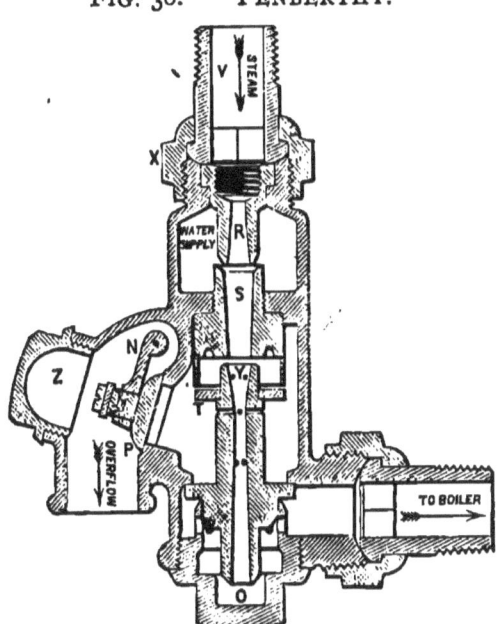

FIG. 36. "PENBERTHY."

steam jet. The following tests of an injector of this pattern have been furnished by the manufacturers, but were made by disinterested experts—height of lift = 3 feet:

Limiting feed temperature at 65 lbs. steam, 128°, delivery 200°
"   "   "   75   "   121°   "   196°
"   "   "   85   "   117°   "   200°
" back pressure without overflow at 65 lbs., 92 lbs.
"   "   "   "   75 " 103 "
"   "   "   "   85 " 112 "

In conclusion it may be said that the construction of the injector is exceedingly simple, as starting is effected by opening the globe valves placed on the steam and feed pipes.

8

Among other well-known patterns of injectors designed especially for stationary boilers may be mentioned the Hancock Inspirator, Park, Eberman, Jenks, Sherriff, Buffalo, Globe; limit of space prevents the insertion of descriptions and illustrations of these also, and a sufficient variety has been given to elucidate the principles upon which all are designed, rendering repetition unnecessary.

The setting of an injector and attaching the various pipes to a stationary boiler are much more simple than in the case of a locomotive. When required to lift, the chief object is to obtain a position as near the level of the water supply as possible, without sacrificing convenience of handling, as the greatest delivery is thus obtained and less care and precision required for starting and regulating. Longer service will also be given before repairs become necessary, because the greater the tax that is placed upon the lifting powers of an injector, the smaller will be the deviation from the correct proportion of the tube which will affect its working. The rate at which the delivery decreases as the lift is increased is shown by Table V, on page 76, and it is only adjustable and self-adjusting injectors, and some forms of the double-jet, that are exempt from this rule. The suction pipe should therefore be as short and direct as possible, and carefully tested under steam pressure to detect any faulty joints or leaks in the pipe; if very long, it should be made at least one size larger than the nipple of the injector, and should be sloped on long horizontal stretches so as to drain toward the well without air or water pockets. Foot valves are not recommended, but strainers should be used, having entrance holes smaller than the smallest diameter of the delivery tube, with an aggregate area of at least four times that of the suction pipe.

The steam pipe should be attached to the dome of the boiler, or connected with a main of such size that the pressure of the contained steam will be constant and the same as that in the boiler. This pipe should be covered, and supplied with a pet-cock for draining accumulated water, espe-

APPLICATIONS OF THE INJECTOR. 121

cially if the conditions under which the injector must operate are at all adverse. Every effort should be made to supply as dry steam as possible, and this can only be obtained by carrying out these recommendations.

The delivery pipe should of course be short and direct, although this is not as necessary as with the other connections; it should be supplied with a main check valve, a stop valve and a pet cock for draining. The vertical length of this pipe or the position of the injector relative to the water level of the boiler is comparatively unimportant, as there is always an excess of power in the jet more than sufficient to overcome any ordinary difference in level. In the tests of the Penberthy Injector there is shown to be an excess of counter pressure over the initial of 28 pounds at 75 pounds steam, permitting a difference in level and friction head of about 56 feet, far greater than is likely to occur in practice.

All steam and water pipes should be carefully blown out with steam to free them from dirt and scale before the injector is set in place, and all pipes should be bent so that the unions will fit squarely in their seats, requiring no force to draw them to place when the coupling nuts are screwed up.

It may appear that needless stress has been laid upon the method of attaching injectors, and that unnecessary refinements have been suggested, yet it may be said that the advantages arising from good work and conscientious attention to details of this kind more than compensate for the additional outlay. In this respect, English engineers are much more particular than those in this country; the general foreign practice is to substitute pipes bent in easy curves for all cast iron elbows, flanges take the place of malleable iron couplings, while the general solidity and care manifested in the attachment of boiler fittings are well worthy of imitation.

BOILER TESTERS.

The principle of the injector has been applied very satisfactorily to the testing of boiler seams, and special instruments are made for this purpose. They are so constructed that the boiler may be rapidly filled with warm water and high testing pressure then applied, even though the jet is actuated by low pressure steam.

In this instrument, two sets of tubes are used, one for each operation; the primary set acts as an ejector, delivering between 600 and 700 cubic feet of water per hour, and fills the boiler up to the safety valve; all openings are then closed, and the smaller set, having a capacity of about 60 cubic feet per hour, is started, and the pressure in the boiler raised to the point desired; this secondary set is proportioned like an injector, but has a much larger steam nozzle than would be ordinarily employed, as the excess of counter pressure required is much greater than in the case of an injector. If the same relations that obtain in the exhaust injector would hold true at 70 pounds steam, testing pressures of 450 pounds would be produced, for with an absolute steam pressure of 14.69 pounds, 105 pounds, or seven times the initial, can be obtained; but in this case the area of the steam nozzle is nearly sixteen times that of its delivery tube, while at 60 to 80 pounds, 4½ is the greatest ratio that can be used efficiently. The usual pressures are about as follows:

30 pounds initial will give 135 pounds terminal pressure.
40   "   "   180   "   "
60   "   "   260   "   "
80   "   "   320   "   "

This method of testing is very convenient for boiler-shop use, as the device is small and compact, easily applied, and requires little skill to operate. Safety valves, which usually form part of the apparatus, may be set for any desired pressure up to the limits of the instrument, so that there may be no danger of subjecting the boiler to excessive strain. The feed supply should flow under an unvarying head so that the conditions under which the jet operates may remain constant, and rather higher counter pressures are given when the water is received under a head than when it is lifted.

The suggestion has also been made to apply the power of the injector for increasing available pressure, to use in a

APPLICATIONS OF THE INJECTOR. 123

hydraulic press, but it is doubtful if this would ever be of much practical value. By properly proportioning the piston areas of the press, an injector could be substituted for the pumps usually employed, but practical considerations would at once suggest themselves which would render the superiority of the injector exceedingly dubious.

For fire pumps there is a field that is more promising. Small instruments are constructed having steam nozzles about the same size as the delivery tubes, and are frequently attached to yard and switching locomotives with good results; one of the most prominent railroads in this country has supplied its yard engines with pumps of this kind, which are operated by simply opening the tank valve and the steam valve, and coupling up a line of hose. Fig. 37 gives an exterior view of this style of fire pump which will throw 4600 gallons of water per hour through a $\tfrac{11}{16}$ fire nozzle a distance of 115 feet. It can be coupled to an ordinary fire hydrant in case the supply of water in the tank is insufficient. The capacity and distance given are based upon a steam pressure of 125 pounds per square inch.

FIG. 37. FIRE PUMP.

For draining or raising large quantities of water a small height, ejectors can be used with a comparatively small expenditure of steam, although with considerable less economy than if a pump were used, and there are but few cases where the increase in temperature— the chief source of economy in the injector as a boiler feeder —would be advantageous. The advantage of steam condensation is of use indirectly in the case of bilge pumps, which, beside performing the function for which they are designed, are often operated when the engines of steamers are stopped at sea as condensers for the sudden accumulation of steam until the fires can be properly checked. These pumps are simply constructed, and, having no valves, are not liable to stoppage.

The proportions of the parts of an ejector vary somewhat with steam and counter pressure and the purpose for which it is to be used; the ratio of the area of the steam orifice to the delivery tube may vary from 0.25 to 1.00, to 0.8 to 1.00, the former giving a strong suction in a well-shaped tube at the higher steam pressures.

As exhaust steam condensers for reducing the back pressure upon the piston of an engine, special forms have been designed, which are complete in every detail for regulating the water supply for varying loads, and large numbers are now in use. The arrangement is compact and reliable, requires no air-pump, and is usually less expensive than the other systems.

There are numerous other devices in which jets of steam are used for moving and forcing air or other gases, such as exhausters, blowers, ventilators, etc.; but as they do not depend upon the condensation of the steam which is in reality the inherent characteristic of the injector, they do not come within scope of these pages.

CHAPTER IX.

DETERMINATION OF SIZE—INJECTOR TESTS—DIAGRAMS OF RESULTS.

As has already been shown, the maximum capacity of an injector depends upon three factors: Steam pressure, height of lift, and temperature of feed water. It is upon the design of the tubes and the special type of the instrument that the extent of the variation depends when any two conditions are given and the third altered. This has been illustrated by the approximate formula, No. 23, on page 79, which shows the rise in capacity with the pressure; by Table VII, page 78, where the steam pressure and the lift remained constant, and the feed temperature varied; and Table V, page 76, where the pressure and temperature were constant, and the height of lift increased. Two diagrams, Figs. 39 and 40, indicate graphically the first and second cases more clearly than can be done by tabulation, showing the general rules that hold true, though in a varying degree, for all styles and patterns of injectors. It is thus seen that judgment—or, preferably, experiment—must be used to determine the capacity of an injector for any conditions differing from the usual practice, and a full statement of the conditions should be submitted to a competent manufacturer whose wider experience will enable him to give accurate information.

For each given feed temperature and height of lift there is a limiting steam pressure, beyond which there is no further increase in the capacity; this limit of pressure is lower with injectors designed for stationary boilers than those used in locomotive service, and it is usually a function of the inlet area to the combining tube, and varies with the pattern of injector; considering these things, it is always better before deciding on the size required to submit the following data to the manufacturer:

(a) Required capacity in cubic feet per hour,
(b) Steam pressure,
(c) Height of lift,
(d) Temperature of feed water.

Stating, also, any special conditions, such as length of suction pipe, if unusual, or an excess of back pressure, often required when feeding through a heater. Some forms of injectors are so designed that all ordinary conditions of service are satisfactorily covered by the usual form supplied; but others require slight modifications of the forms of tubes which enable them to operate more effectively in special cases.

The first information to obtain is the maximum evaporation of the boiler; formerly the allowance granted per nominal horse power was one cubic foot of water; but with the great advance that has been made in the economy of the modern stationary engine, about one-half that amount, or 35 pounds per horse power, is usually regarded as sufficient. It is always more advisable to obtain the actual evaporation, but when unknown or doubtful, the requirements may be obtained from the estimated steam consumption of the engine, calculated from an indicator card, allowing sufficient margin for other drains upon the steam supply. If this cannot be done, the approximate rules here given can be used:

To FIND WEIGHT OF WATER EVAPORATED PER HOUR $(=X)$ FROM AND AT 212°.

When the heating surface $(=H)$ is known:

For tubular boilers, $X = H \times 3.40$
" flue " $X = H \times 6.00$
Water tube " $X = H \times 3.5$.

When the grate surface $(=G)$ is known:

For externally fired boilers, $X = G \times 94.00$.
" internally " $X = G \times 180.00$.

When the total weight of combustible burned per hour $(=C)$ is known:

$X = C \times 9.00$, (Assumed as a fair average.)

SIZES OF INJECTORS.

When the heating surface (H) and grate surface (G) are known:
For internally fired boilers, $X = 36.0 \sqrt{H \times G}$.
(Molesworth Pocket Book.)
When the heating surface (H) and weight of combustible (C) are known:

$$X = \left(12 - 4.375 \times \frac{C}{H}\right) \times C.$$

NOTE.—To reduce weight of water evaporated from and at 212° to the corresponding evaporation from a feed temperature of 70° and a steam pressure of 70 pounds, divide by 1.183.

To find weight of combustible, subtract from total weight of coal burned per hour, the weight of the ashes and the contained moisture.

In determining the size of injector required for a locomotive, the size of the cylinder is usually taken as the standard, although the diameter of the boiler, and the special service for which the locomotive is intended, has a modifying influence. The following table embodies the practice of the Baldwin Locomotive Works for all engines equipped with two injectors, one injector and one pump, one injector and two pumps:

TABLE IX.

SIZES OF INJECTORS FOR LOCOMOTIVES.

Diameter of Cylinders. Inches.	Nominal Size of Injector.		Diameter of Cylinders. Inches.	Nominal Size of Injector.	
	2 Injectors, or 1 Pump and 1 Injector.	2 Pumps and 1 Injector.		2 Injectors, or 1 Pump and 1 Injector.	2 Pumps and 1 Injector.
6, 7, 8	3	3	19	8*	6
9, 10	4	3	20	9	6
11, 12, 13	5	4	21	9*	—
14, 15	6	4	22	10	—
16, 17	7	5	23	10	—
18	8	6	24	11	—

* Use next size larger with specially large boilers.

Almost all of the best known patterns of locomotive injectors of the same nominal size correspond closely in maximum capacity, even though the weight of steam used per hour will vary according to the proportion of the nozzles and the general excellence of the design. With injectors for stationary boilers there is less uniformity, and no general standard prevails.

With instruments purchased from reputable manufacturers, there is but little danger of the capacity falling below the figures claimed if the injector is operated under the conditions prescribed in the catalogue; owing to the present refined methods of duplication of parts, nozzles can be bored, reamed and turned with mathematical accuracy and endless repetition, so that there is no excuse for poor workmanship or any deviation from proper dimensions; but there may be flaws or spongy places in the main castings, and these defects do not always become apparent until after the injector has been subjected to the vibration of an engine, escaping detection even in the proving room of the manufacturer. If an imperfection of this kind should occur in one of the walls separating the steam from the feed chambers, a useless heating of the feed water would result, and both the maximum capacity and the highest admissible temperature of the feed water would be reduced. Assuming the injector to be in perfect condition, there may be a very serious falling off in the capacity below that claimed, if the steam pressure or the height of lift during the test are greater than those used by the makers; for instance, if the standard were based upon a pressure of 120 pounds per square inch and a lift of 1 foot, the capacity (based upon an actual case) might fall off 10 per cent. when the lift is increased to 6 feet, and 25 per cent. when the steam pressure is raised to 150 pounds.

These considerations show the advantage, and, in some cases, the necessity of having a thoroughly-equipped testing department wherever large numbers of injectors are in use, both for the purpose of experiment, but more especially for proving those that have been repaired by the employee in charge of this work. The exact conditions occurring in practice—steam pressure, height of lift and length of suction pipe—should be imitated as nearly as possible. By this system the performance of all injectors submitted may be carefully determined without premature acceptance of manufacturers' claims. Many of the large railroad companies have

SIZES OF INJECTORS.   129

appreciated these facts, and installed in their laboratories apparatus for testing new forms of injectors, as well as to prove one or two selected from all lots submitted for regular service, which must conform to the standard required by the railroad.

The standard dimensions to which all injectors used on the Pennsylvania Railroad are required to conform are shown in Fig. 38. and cover chiefly the position of the steam, feed and suction pipes, and the attaching bolt, so that all patterns of injectors may be interchangeable. As these proportions are known as the P. R. R. Standard, and are used by several of the large companies, a list of the principal dimensions is given in the table accompanying Fig. 38.

FIG. 38,

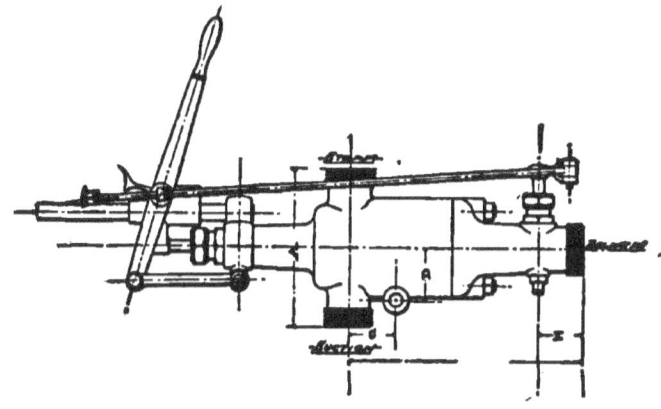

Nominal Size of Injector.	A	B	C	D	E
No. 6	$7\frac{1}{2}$	$11\frac{3}{8}$	$2\frac{1}{2}$	$2\frac{11}{16}$	$2\frac{1}{8}$
No. 7	$7\frac{1}{2}$	$11\frac{3}{4}$	$2\frac{1}{2}$	$2\frac{11}{16}$	$2\frac{1}{8}$
No. 8	$8\frac{7}{8}$	$14\frac{3}{4}$	$2\frac{7}{8}$	$3\frac{1}{16}$	$2\frac{7}{8}$
No. 9	$8\frac{7}{8}$	$14\frac{3}{8}$	$2\frac{7}{8}$	$3\frac{1}{16}$	$2\frac{7}{8}$

Regarding the method of testing, it may be said that the simplest way is to empty a tank of known dimensions between given levels, delivering against a check or lever valve loaded to the initial steam pressure. A better method is to weigh the feed water, and if possible, the delivery also, so that the weight of steam used per hour can be determined; with small injectors this can almost always be easily done, but with locomotive sizes it is hardly practicable to weigh even the feed. From an experimental point of view it is interesting to have both weights, but the proportion of water to steam can be easily determined by observing the feed and delivery temperatures, and applying formula (33) given on page 83, which is sufficiently accurate for all ordinary purposes; the measuring of the temperatures is best done by inserting the thermometer bulbs directly in the flow of the water, screwing the body through a tee having a ¾ gas pipe thread; special instruments are made for this purpose, and are more satisfactory than the inserted mercury cup. The use of a spring check valve in the delivery pipe is seldom satisfactory, and does not give as accurate results as discharging directly into the boiler, or, what is more practicable, delivering into a steam trap having a large balanced outlet valve, and connected to the steam area of the boiler. This arrangement is decidedly the best, as it does not alter the level of the water in the boiler, and the pressure against which the injector delivers is maintained constant, giving more precise results when the minimum capacity and the highest admissible temperature of feed water are to be determined.

The most convenient and instructive method of recording comparative tests is to plot the results in the form of the diagrams given in Figs. 39 and 40. Here are shown tests of three of the best known locomotive injectors in this country, the object being to ascertain the increase in capacity with the steam pressure, and the effect of high feed temperatures upon the maximum and the minimum capacities. In both diagrams the lines parallel with the base line represent gallons per hour; those at right angles indicate steam pressure in Fig. 39, and feed temperature in Fig. 40.

These series of tests were made in the laboratories of two of the large railroad companies at considerable time and expense,

FIG. 39.
COMPARATIVE TESTS OF THE MAXIMUM AND MINIMUM CAPACITIES OF THREE PATTERNS OF NO. 9 INJECTORS AT VARIOUS STEAM PRESSURES. Feed, 74°; lift, 6 feet.

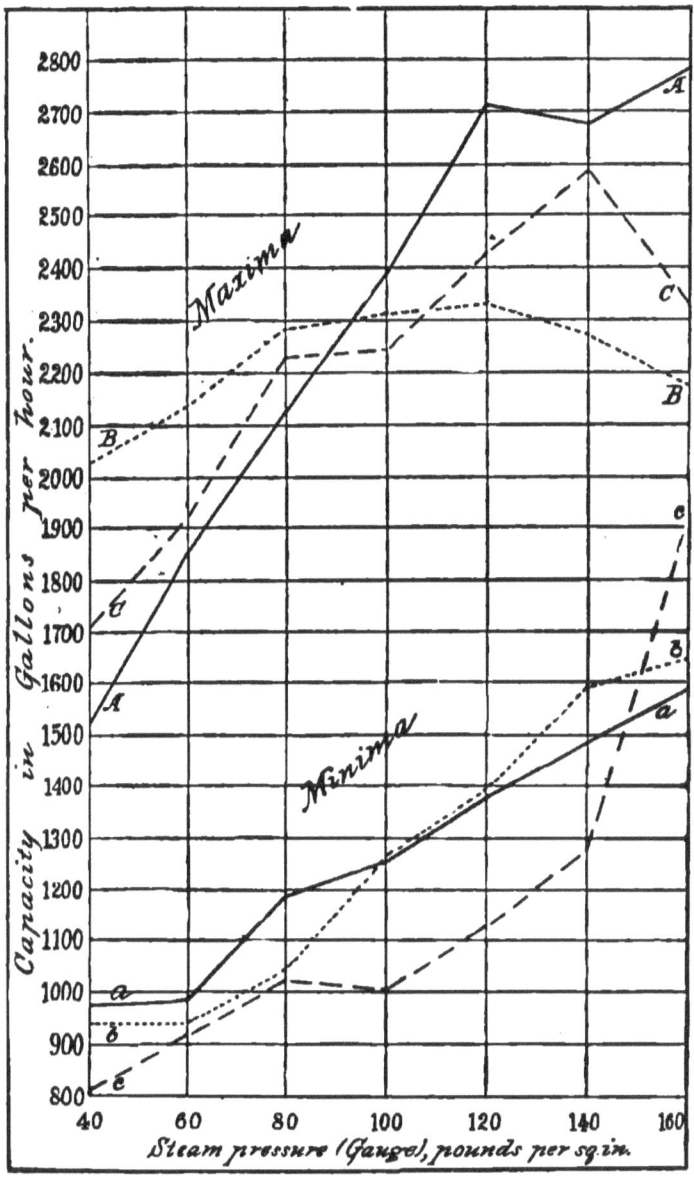

and are entirely unbiased; yet, in order to avoid any comparison of the relative merits of the different patterns, they are here indicated alphabetically, the same letters applying to similar injectors in both diagrams, even though the sizes are different. The maximum capacities are designated by capital letters, and the minima by small letters. The injector marked C^1 is not the same pattern as the C shown in Fig. 39; but as it is the product of the same manufacturer, and has similar characteristics, it was added to the diagram.

The capacities in Fig. 39 were obtained by noting the time required to deliver a measured weight of water against a spring check valve. The lines connecting the observations should form symmetrical curves, and all irregularities indicate possible errors of adjustment or of observation which are difficult to detect unless the results are plotted; some patterns of injectors require very careful adjustment of both water and steam, and slight variations in the regulation materially affect the result. The two lines, A and C, keep rather close together, while B delivers more water at the lower pressures, and less beyond 120 pounds, at which pressure it reaches its maximum. A apparently starts downward between 120 and 140 pounds and then ascends, indicating imperfect regulation at the reëntrant angle. Referring to the minima, the lines a and b almost coincide, while c gives a much better minimum until 160 pounds is reached, when the sudden rise indicates insufficient throttling of the feed valve. If the steam pressure were raised still higher, a point would be found where the converging lines of the maxima and minima would intersect, which would be at the highest steam pressure with which each injector can be operated under these conditions.

In the tests given in diagram, Fig. 40, the steam pressure and the height of lift are constant, and the feed temperature raised 5° after each observation until the injector broke. This was done to obtain under each condition the maximum and the minimum capacity of each injector; the latter lines have numerous reëntrant angles, due to the fixed notches which the regulating mechanism necessitated. In all cases

TESTS. 133

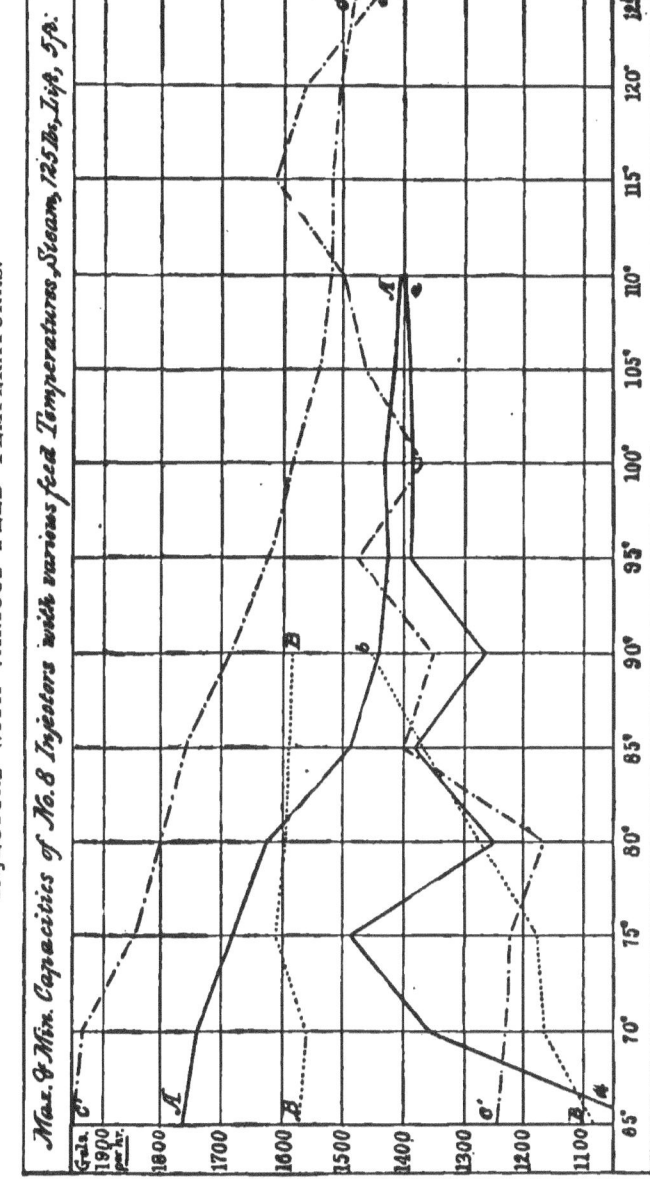

FIG. 40. COMPARATIVE TESTS OF MAXIMA AND MINIMA OF THREE PATTERNS OF INJECTORS WITH VARIOUS FEED TEMPERATURES.

the two capacities should be identical at the breaking and overflowing temperatures; this is true with lines A and C, and probably with B also, if prolonged to 94°. The termination of each line indicates the highest admissible temperature of the feed water for each pattern of injector, A breaking at 110°, B at 90°, and C at 125°.

A rather curious feature of the tests with the $C^1 c^1$ injector is the crossing of the two lines, showing an apparent paradox, that the minimum capacity at 115° is higher than the maximum. In this pattern the regulation was obtained by reducing the steam supply, and as it is well known that with high pressures and temperatures a smaller steam nozzle can be used to advantage, it follows that here, where the steam area is reduced, the capacity is larger than when the full steam nozzle is used, but which, when fully opened, gives best results at lower temperatures.

By contrasting these results, will be seen the difficulty of obtaining a formula which will cover accurately all the conditions that enter into the performance of an injector, as the actual construction exercises such a modifying influence upon the results. The equations which were given were based upon the form which seemed to approximate the theoretical results, which would of course be a self-adjusting injector able to preserve the condition of the jet at the instant of approaching the smallest diameter of the delivery tube nearly at a density of unity. In all experimental work upon which theoretical conclusions are to be based, great care should be exercised to secure accurate observations, and these should at once be plotted in the manner shown, so that all doubtful results can be checked, and, if necessary, repeated.

A more recent test than those already described, and one in which special precaution was taken to insure accuracy, will now be given; this test* was made by disinterested experts with apparatus that can be duplicated in almost any railroad shop with but little preparation. Any official desiring to make a comparative test of the injectors in use on his line, can apply the method outlined with satisfactory

* Reprinted by permission from the *Railroad Gazette*.

and decisive results at comparatively small expenditure. The injector used in this test was a No. 10½ of the Improved Sellers 1887 pattern, and it was chosen in common with several other well-known injectors for the purpose of making a selection for the equipment of a large number of locomotives, and to ascertain if the performance at certain given steam pressures fulfilled the requirements of the specifications.

FIG. 42.

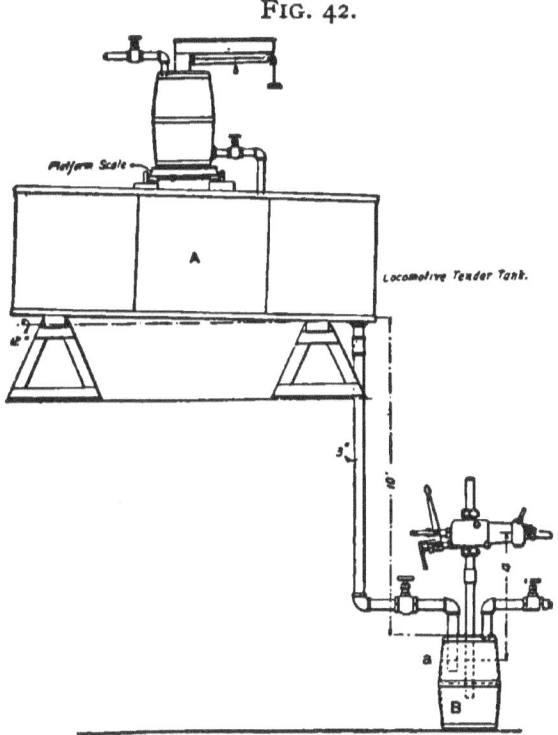

Apparatus.—The injector (see page 104) was supplied with dry steam from a 200 H. P. Babcock and Wilcox boiler, through a 3 in. pipe carefully lagged with asbestos covering. The injector was bolted against the side-wall of the boiler with the starting lever and water supply valve within convenient reach of the operator. The water supply was maintained at a constant level in a large barrel directly below the injector, into which the suction pipe was extended to within one foot of the bottom; this pipe was 2½ in. diameter up to the nipple of the injector, where it was reduced to 2 in.

The water supply was weighed and delivered into the suction barrel as follows: Ten feet above the level of the suction barrel were two large tanks, forming a reservoir capable of holding about 1100 gals. From the 3½ in. flanged pipe bolted to the bottom of each was a vertical 3 in. pipe extending into the suction barrel, 6 in. below the 4 ft. level; this pipe was made amply large, so that under the head available the tanks would be drained and the pipe emptied before the level of the water in the barrel could be lowered by the injector more than 6 in.; owing to the slope of the bottom of the tanks and the large size of the emptying pipe, this was accomplished without the slightest trouble. A globe valve was placed in this pipe close to the injector, so that the proper level of the water in the suction barrel could be maintained by the operator. (See Fig. 42.)

Above and resting on these water tanks was a 500 lb. platform scale carrying a 47 gal. barrel, into the bottom of which was screwed a 2 in. outlet pipe for emptying it quickly into the iron tanks below; this pipe was kept clear of the sides of the tank and the scale. Another pipe brought the water supply from the city mains above the top of the barrel, and as it was supplied with a quick acting, tight gate valve, the barrel could be quickly weighed empty, filled, re-weighed and emptied into the iron tanks until the quantity of water required for a run of from 15 to 20 minutes was obtained.

Additional means for supplying the barrel with water during the preliminary run before each experiment was found to be necessary, because when the reservoir was filled with weighed water, none could be withdrawn until the moment the experiment commenced. For this purpose a 2 in. hose was run into the barrel and a valve placed near the discharge end, and the water level maintained constant until all preparations were completed; this valve was closed and hose withdrawn before the valve for the reservoir tanks was opened.

Gauges were placed so that the pressure in the steam pipe or in the delivery pipe could be obtained alternately on either

TESTS. 187

FIG. 43. ARRANGEMENT OF PIPING AND GAUGES FOR TESTING INJECTORS.

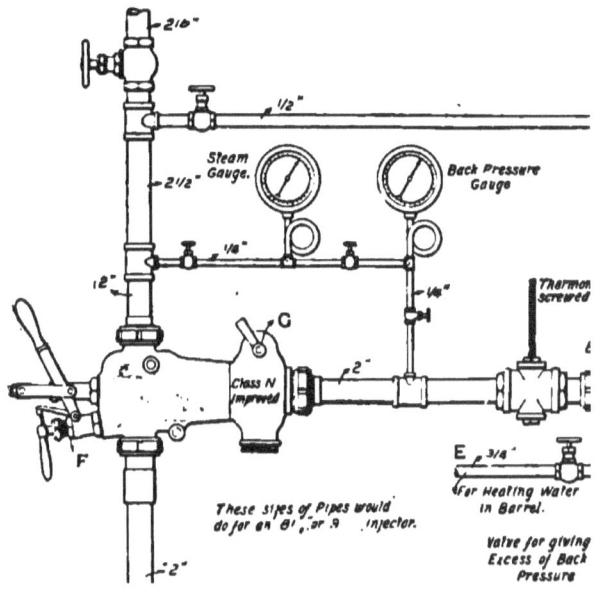

THE GIFFARD INJECTOR.

or both gauges, or simultaneously on separate gauges; this arrangement worked very satisfactorily, and the admission of any error due to the difference between the steam and delivery pressures, or to a discrepancy between the two gauges, could be prevented. Fig. 43 shows the arrangement of pipes, valves and gauges by which this was accomplished, and is self-explanatory.

As the delivery of the injector was too great to be taken into the boiler without affecting the steam pressure carried, it was passed through a special balanced valve (Fig. 44), which maintained a constant pressure equal to that of the boiler. The delivery pipe of the injector was coupled to the under side of a check valve, which was connected to a piston

FIG. 44.

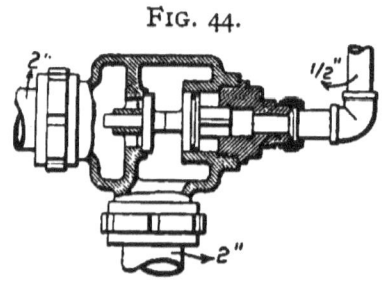

of the same area, upon the upper side of which full boiler pressure was obtained by a pipe tapped into the steam supply.

Measuring Devices.—The steam gauges had been subjected to careful test and calibration by the makers; at the same pressures the readings of the two gauges agreed exactly.

The thermometers were tested in oil every five degrees, both up and down, and the corrections noted, from which a table was made and used to obtain actual temperatures. The delivery thermometer was brass cased, and was screwed into the delivery pipe close to the injector, with the bulb well immersed in the passing water; the scale was divided to single degrees, and could be read easily to half degrees. It was corrected for the error due to the compression of the bulb, as it was subjected to the pressure of the delivery.

The scale used for weighing the feed water was tested by United States standard weights and found to be correct. The barrel resting upon the scale platform was weighed before and after each filling, so that the exact net weight of water passing into the receiving tanks each time could be determined.

The filling of the water tank reservoir required two observers, one to open and close the inlet valve from the city mains and the valve from the bottom of barrel leading into the reservoir, and the second to shift the tare weight for the empty and the full barrel, and to record the weights upon suitably prepared blanks. As the scale registered to $\frac{1}{4}$ lb., the possible error was very small, for a test by United States standard 100 lb. weights after the experiment was completed, showed no change,

Method of Testing.—The reservoir tank having been filled with the required weight of water, the valves in the steam pipe were opened wide, and the water drained out; the water regulating valve on the injector was opened and the cam lever over the waste valve was set so as to allow this valve to open freely. Precaution was taken to insure the water supply being free from dirt and chips and the suction barrel clean. The 2 in. hose from the water main was led into the barrel and the injector started against full back pressure. The hose discharged beneath the surface of the water to prevent air being carried down into the water and interfering with the free flow in the suction pipe. Observations were made as to the regularity of the steam pressure, and the readings of the steam gauges and the thermometers in the suction barrel and delivrey pipe were found to be practically constant; an observer was stationed at the water valve in the 3 in. pipe leading from the overhead reservoirs, another to read the gauges and thermometers, and a third to take and record all readings and note the general performance of the injector. When everything was ready, the barrel was rapidly filled through the hose and then its valve closed and the hose entirely withdrawn; as soon as the water level in the barrel was drawn down by the injector to the lower white line (*b*) (see Fig. 43), the recorder noted the exact time on a stop watch, the other observers noted the thermometer and gauge readings, while the valve in the 3 in. feed pipe from the reservoirs was quickly opened and the water level raised to the line (*a*) four feet below the centre of the injector, where it was maintained during the continuation of the experiment by careful regulation of the valve. Readings of the thermometers and the gauges were taken every three minutes until the reservoir was empty, which could be immediately noted by the rapid falling of the level of the water in the barrel; just before the lower level was reached the end of the 3 inch pipe from the reservoir was exposed, this construction being insisted upon so that the observer could be certain that all the water in the reservoir had flowed into the barrel; when the lower level—from which the start was made—was reached the signal was given to the recorder, and the time again noted, the difference in time being that required to lift and force against initial pressure the total weight of water contained in the reservoir; from this could be calculated the capacity of the injector in pounds, cubic feet or gallons per hour.

The method of determining the minimum was the same, except that occasional adjustment of the regulating valve was required during the experiment owing to variations in the pressure of the steam; also, the quantity of water weighed into the reservoir was less than half that used for determining the maximum. Care was taken that the counter pressure produced by the back-pressure valve should always be equal to that of the boiler so as to obtain precise results.

From these two sets of experiments were determined the figures for the ratio of the minimum capacity to the maximum; subtracting this from 100 gives the "range" in percentage of maximum.

To determine the relation between the weights of the supply water and of the steam required to force it into the boiler, it is evident that the simplest method would be to subtract the known weight of the supply from the weight of the delivered water and then divide the weight of the supply by this difference. With small injectors this is often done, as the volume of water to be handled is not large, but with an instrument of the size used, this method becomes impracticable. The method of delivery temperatures was therefore substituted and the same results obtained without the necessity for weighing the delivery. The formula used was the following: (See page 83.)

$$W = \frac{H + 32 - (.003)\ P - T}{T - t + (.003)\ P}$$

$W =$ Weight of water delivered per pound of steam.
$H =$ Total heat in one pound of steam (absolute) pressure above 32 deg. taken from steam tables.
$T =$ Temperature of the delivered water.
$t =$ Temperature of the water supply.
$P =$ Steam pressure (gauge).

Maximum water-supply temperatures were obtained by returning some of the water from the delivery pipe to the barrel or reservoir; care was taken that the hot and cold water should be thoroughly mixed and that the temperature should not be increased too rapidly. Two sets of results are given: limiting temperatures at each steam pressure for automatic restarting without subsequent waste of hot water or steam from the overflow; also maximum operating temperature at which the injector will run without the jet breaking; the former were obtained with the waste valve free to rise on its seat; the latter, with the waste valve closed by throwing the cam lever backward; in this case if the jet breaks, steam will flow back into the suction pipe until the waste valve is allowed to open or the steam supply is shut off.

Numerous special tests were also made to determine the action of the injector under conditions frequently occurring in practice, such as variations of the steam pressure, hot water in suction pipe, and the effect of a temporary interruption of the water supply, such as would occur when the movement of the water in the tank of a locomotive uncovered the end of the suction feed pipe; also, the amount of water wasted during starting and stopping. An account of these tests will be made under the heading of "Results."

It should be noted that all the experiments were made without throttling the steam supply; this was found to be necessary as an early experiment at 150 lbs. steam showed that the superheating due to wire drawing materially affected the results; in all subsequent tests the pressure of the boiler was raised or lowered to meet the requirements of the experiment.

The directions given in the catalogue of the manufacturers for stopping and starting the injector were followed: To start: Pull out the lever. To stop: Push in the lever. Regulate for quantity with water valve. In starting on high lifts and in lifting hot water, it is best to pull the lever slowly.

Results.—To facilitate the comparison, the performance of the injector at different pressures and the results obtained at each set of experiments have been plotted in separate diagrams, forming curved or broken lines connecting the several observations, so that the results for any intermediate condition can be easily determined; as the scale of the diagrams is necessarily small, a complete table of results has been given, which contains the actual figures obtained. The results of the tests were remarkably good, for in several cases the claims of the manufacturers were much exceeded. Accompanying each diagram is a short review of the results.

As stated above, however, certain tests were made which could not be tabulated, but are almost equally valuable in considering the general performance of a locomotive boiler feeder.

(A) Variation in steam pressure. The injector was started with the lever-starting valve and the water-regulating valve wide open, and the pressure in the boiler and the back pressure were simultaneously lowered from 200 lbs. to 120 lbs. and then later, with all valves as before, from 120 lbs. to 40 lbs. steam, without a drop of water appearing at the overflow; raising the same pressure caused no overflow of steam or water at the waste pipe, and the injector seemed to operate as successfully at one pressure as another, without making change in the tubes or in the position of its steam or water valve.

(B) From the fact that this injector worked very satisfactorily with hot supply water, it was evident that its lifting power with the suction pipe warm, would also be good; owing to the provision of large overflows in the forcing combining tube, it is not necessary that care should be used in admitting steam to the main jet after priming—as is the case with other forms of injectors—for even though the feed water be above the limiting temperature as it comes from the lifting nozzle, the forcing jet will not break, but will cause an overflow of steam and hot water until the hot water is drawn out, which usually occurs in a few seconds; in this case the amount of waste was small, but with cold water only a few drops appeared at the waste nozzle at 60, 120, 200 lbs. of steam or intermediate pressures. The mean of a number of tests, stopping and starting the injector with the supply water at ordinary temperatures, gave one-half pint as the amount wasted each time.

(C) It has been found by the experimenters that the effect of admission of air to the suction pipe of all injectors which adjust their capacity to suit variations in the steam pressure is to immediately break the jet, and to cause the steam to blow back through the hose into the tank; but with this instrument any such interference with the normal condition of the jet causes a waste of steam or water at the overflow pipe, which ceases as soon as the disturbing cause is removed. To test this feature, the water in the suction barrel was allowed to fall below the lower end of the suction pipe, so that air would be sucked up into the injector. This caused a discharge of steam and air from the waste pipe, which ceased as soon as the usual level of the water in the barrel was restored. This test was repeated at 200 lbs. steam pressure, when the lifting of the supply water and the forcing it into the boiler occurred the instant the water covered the end of the suction pipe.

In regard to the injector itself, it may be said that it responded promptly at all times to the movement of the starting valve. It is started and stopped by the continuous motion of a single lever, and was regulated by a side motion of the quadrant regulating lever only for the purpose of altering the amount of delivery. Its construction is simple and easily understood; no outside rods, levers or bell-cranks are used, nor complicated internal valves. When hot water is to be lifted it was found that the strongest suction was obtained when the starting lever was drawn forward about 1 in. and

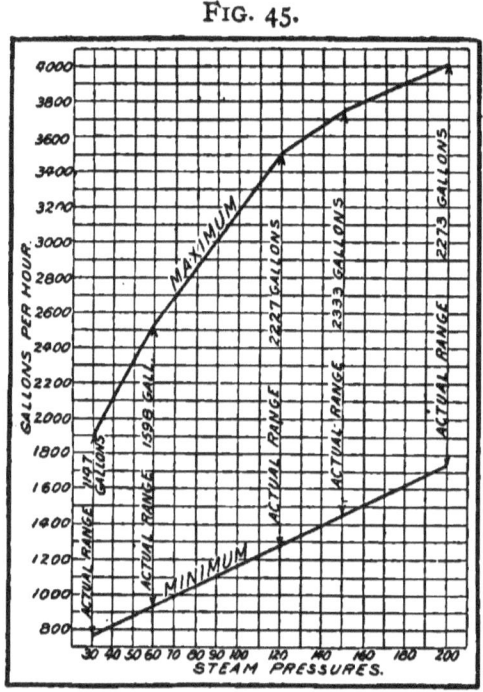

FIG. 45.

the remainder of the stroke given after water appeared at the waste pipe.

The Diagrams.—Fig. No. 45 is a graphical representation of the capacity tests given in lines 1 and 5 of the Table of Results. At the top, ranging from left to right, are gauge pressures in pounds per square inch, from 30 to 200 lbs.; the horizontal lines indicate gallons of water taken from the supply tank per hour, so that the heavy curved line shows the change in the capacity with the steam pressure. The maximum capacity increases as the pressure rises from 1,912 gals. at 30 lbs. to 2,535 at 60 lbs., then 3,510 at 120, 3,760 at 150 and 4,000 at 200 lbs. steam; the last capacity is higher than that at any lower steam pressure, and was above that of any other injector of the same size, even though the capacity may be the same at 120 or 150 lbs. The minimum is shown by the lower heavy line of the diagram, and increases from 765 gals. at 30 lbs. to 1,732 at 200 lbs. steam; this possible reduction of the capacity at 200 lbs., from 4,005 gals. to 1,732, is such a great variation in the amount of water deliv-

FIG. 46.

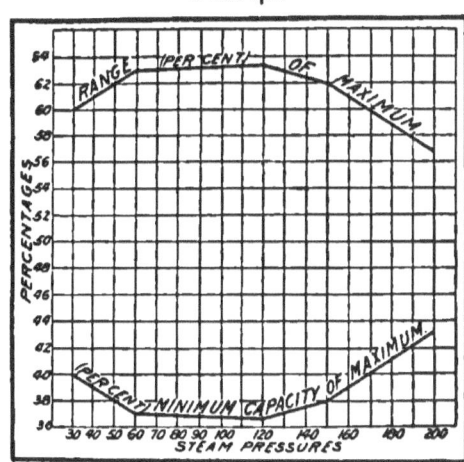

ered that it is very evident that the injector can be used to feed the boiler continuously with either a light or heavy train.

The ratios between the maximum and the minimum are given by the heavy line in the lower part of Fig. No. 46, and the range in percentages is given in the upper part. These same values are given in lines 8 and 10 of the table.

Upon the figures obtained during the test to determine the maximum capacity, are based the values given in Fig. 47, page 146, also line 4 of table, which show the weight of water taken from the supply tank per lb. of steam used by the injector. This represents the actual amount of mechanical work done by the steam, and is a point of special value to practical men; it is a good gauge of the efficiency of the design of the injector and of its economy as a boiler feeder, as it indicates that the minimum amount of steam is used to perform the work of feeding, and no excess is condensed and utilized only for heating the feed water. Under the same conditions of supply temperature and lift, the weight of water

FIG. 47.

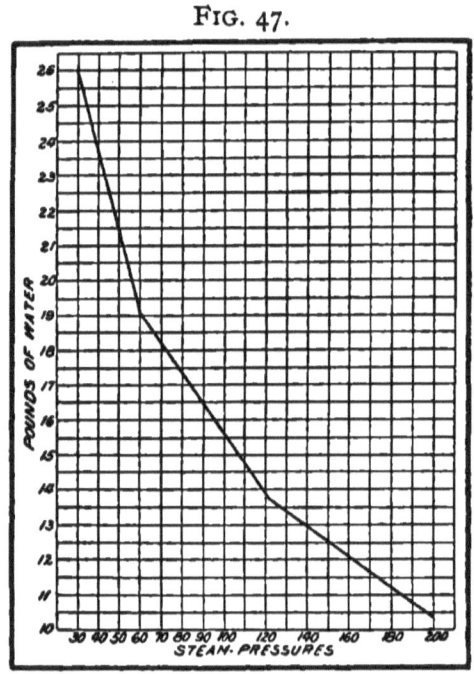

delivered per pound of steam must always decrease as the pressure rises.

A critical comparison of the results obtained on this experiment proves the superior design of this pattern. Referring to the actual figures of Fig. 47 or the table, it is seen that at 120 lbs. steam, 13.6 lbs. of water are taken from the tank and forced into the boiler by 1 lb. of steam and at 200 lbs. 10.34 lbs., the latter result being especially remarkable.

It is very seldom necessary that a locomotive injector is required to feed when the temperature of the supply exceeds 100 degs., but when the occasion demands, the action should be certain and permit a fair range of capacities. Many injectors will not operate at the higher pressures with the supply at this temperature, consequently their action when starting with hot feed pipes causes a very large amount of overflow before the jet enters the boiler. Several temperature tests were made at 30, 60, 120 lbs. steam, etc., and the results are given in Fig. No. 48 and lines 11 and 12 of table. The

FIG. 48.

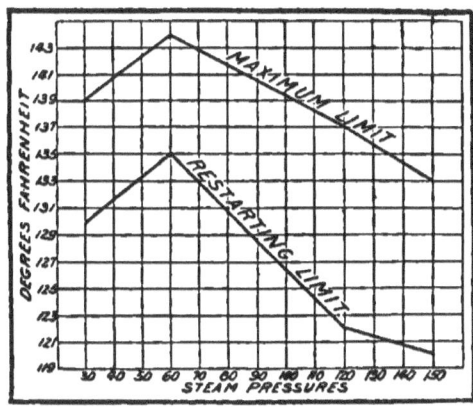

limiting temperature at which the injector is re-started at various steam pressures is given by the lower line, and the maximum temperature to which the supply may be raised before the injector ceases to operate is shown by upper line. The vertical lines of the diagram indicate steam pressures as before and the horizontal lines degrees Fahrenheit. Starting at 30 lbs. steam, the limiting temperature is 138 degs., which rises to 143 degs. at 60 lbs.; at 120 it is 137 deg., and at 150 lbs. pressure 133 degs These were obtained with the waste valve closed to prevent the waste which would occur as these limiting temperatures were approached.

It may be stated that the greatest care was taken to prevent the possibility of error in observation or in accuracy of apparatus. The operators were all skilled in experimental work and the observation and calculation of each were carefully checked. The injector was taken directly from stock and without special preparation, and in performance it exceeded in every detail the requirement of the specification.

TABLE OF RESULTS.

Test of a Sellers' Improved Injector of 1887; Size, $10\frac{1}{2}$; Lift, 4 feet; Supply Water Weighed.

MAXIMUM CAPACITY.

	Mean Steam Pressure	30	60	122	151	200.5
1	Gallons of Water per Hour . .	1912	2535	3517	3765	4005
2	Temperature of Supply Water .	67.0	67.0	54.0	50.0	50.5
3	Temp. of Delivered Water . . .	113.25	125.0	133.4	135.7	154.0
4	Weight of delivered Water per Pound of Steam Used	25.90	19.10	13.60	12.60	10.34

MINIMUM CAPACITY.

	Mean Steam Pressure	30	60	120	148	200.6
5	Gallons of Water per Hour . .	765	937	1290	1432	1732
6	Temperature of Supply Water .	67.0	67.0	54.5	55.0	50.0
7	Temp. of Delivered Water . .	171	212	238	250	263

RANGE.

	Mean Steam Pressure	30	60	121	149.5	200.5
8	Per cent. Capacity of Max. . .	40.0	37.0	36.6	38.0	43.2
9	Actual Range in Gals. per Hour	1147	1598	2227	2333	2273
10	Per cent. Range of Max. Capac'y	60.0	63.0	63.3	62.0	56.8

LIMITING TEMPERATURE OF WATER SUPPLY. DEG. FAHR.

	Mean Steam Pressure	30	60	120	150	
11	Limiting Re-starting Temp. . .	130	135	122	120	
12	Limiting Operating Temp. . .	139	144	137	133	

CHAPTER X.

REQUIREMENTS OF MODERN RAILROAD PRACTICE—REPAIRS AND RENEWALS—METHODS OF FEEDING LOCOMOTIVE BOILERS.

Owing to the introduction of the compound locomotive or the demand for motive power having increased hauling capacity, there has been a general tendency during the last few years, to materially increase the pressure of the steam carried on locomotive boilers, and almost all new engines are designed to carry from 180 to 205 pounds steam pressure.

The designing of an injector for these high pressures is a much more difficult problem than was before presented. The higher the operating pressure, the greater the difficulty in fulfilling the requirements of a good locomotive injector. The criticism against most injectors, and a characteristic of the single jet, fixed nozzle type, is that if it is proportioned to operate at 180 or 200 pounds of steam, it cannot be made to work at the low pressures carried in the round house. There are now in the market several special forms of injector, which show commendable effort to meet the new conditions, and which operate over a wide range of pressure. All injectors with fixed nozzles are most efficient at the special pressure for which the tubes are designed, and although they admit by hand adjustment of considerable variation of the steam pressure, yet at no other higher pressure will they operate as efficiently. With injectors having two sets of tubes, this permissible range is much more extended, but the mechanical efficiency is seldom as high as that of some of the special types.

But the principal feature which interests the user or purchaser of the injector, is the falling off in the capacity as the steam pressure is raised, which often necessitates the purchase of a larger size instrument to obtain the required number of gallons per hour. This entails the consumption of more steam, with a heavier drain on the boiler and consequent loss of pressure. Further, the action at these pressures is less certain and the jet much more sensitive.

Considering the problem in detail, it will be seen that there are a number of features which it is advantageous that a modern locomotive injector should possess, and which are found to a greater or less extent in certain of the most improved types.

One of the chief requirements is simplicity of construction and operation; special care should not be required to prime, regulate and start; every unnecessary movement of the engineer should be dispensed with; the injector should be operated and adjusted by the movement of a single lever, or at most, by two, one for the steam valve and the other for the feed supply. The action should be positive, so that when once started, it can be depended upon to operate continuously without being affected by shock or jar, or change of level of water supply; it is obvious that every demand upon the attention of the engineer by the lubricator, injector or other device connected with the engine, requires time that should be devoted to the operation of the engine itself, watching the track for passing signals or adjusting the cut-off, and therefore each device should be as nearly self-operative as possible.

Fully as important as simplicity of construction, is the general action. A locomotive injector should operate equally well throughout the complete range of boiler pressures without hand adjustment of any kind. Not only is this necessary on the road for the reasons outlined above, but on account of the use of the injector by the hostlers or men in charge of the engines in the round house. It should give the maximum delivery at 200 pounds steam and work with a wide-open lazy cock at 20 and 30 as well as at 200 pounds, for it is evident from the discussion of the theory of the injector, that the capacity should vary with the steam pressure; now if it is necessary to regulate the water supply

REQUIREMENTS.   151

when the pressure falls, for instance, from 180 to 160 pounds, due to a special drain upon the steaming capacity of the boiler, the attention of the engineer is taken from his other duties and must regulate and adjust his feed, to prevent all the delivered water from the injector passing out through the overflow into the ash pan.

It is very desirable also to have a wide range of capacities at the usual working pressure. This is difficult to obtain at 180 and 200 pounds steam, but in order to obtain the best results in feeding, the range should be at least 50 per cent., and more if possible. The special reasons for this will be given in connection with the methods of feeding locomotive boilers.

On page 65 is given the difference between the amount of steam used by two different styles of injectors; this is due in great measure to the use of a larger steam nozzle than is necessary. An injector, is practically a pump in which the actuating steam is condensed. Treat the injector as a mechanical device; that device which gives the required delivery with the minimum consumption of steam or energy, is the most efficient. Every pound of steam taken out of the steam space means a subtraction of a pound of steam from use on the pistons, and there are times during a heavy pull, when steam in the cylinders is much more valuable than hot water in the boiler. The author has known of cases where there has been shown a very material improvement in the performance of a badly steaming engine by changing to a more economical form of injector. By an efficient injector is meant one in which the delivery is large per unit weight of steam. What this ratio should be, is difficult to say, but recent improvements have raised the delivery of water per pound of steam, from 8.8 pounds to 11.2 at 180 pounds pressure. Besides the advantage of having less tendency to pull down the steam pressure, an efficient injector induces a certain amount of economy in the coal consumption, by increasing somewhat the evaporative capacity of the boiler. The greater the difference between the temperature of the waste gases passing into the smoke-box and the feed water, the greater the transfer of heat through the boiler tubes to the water. The temperature of the waste gases varies from 800 to 1200 degrees, and if any part of this waste heat can be absorbed by the feed water, it is a clear gain, and this can be obtained by using a lower temperature of the delivery, with smaller steam consumption.

There is another feature which commends itself to many practical railroad men, not only on account of the added certainty given to the action of a locomotive injector under all conditions, but for the convenience in obtaining the full range of capacities. When an injector is re-starting, it will continue to force water into the boiler as long as it is supplied with water and steam, and will not blow steam back into the suction pipe even if the continuity of the jet should be disturbed by an interruption of the water or steam supply; further, the water supply can be reduced below the actual minimum capacity of the injector without breaking the jet, so that close regulation can be obtained without special care. Operating closed overflow injectors at high pressures on fast express trains is avoided by engineers wherever possible, if there be slightest danger of the injector breaking, for the time occupied in repriming and starting an injector with hot pipes may be two or three minutes, sufficient for the train to have covered as many miles, during which time the attention of the engineer must be more fully occupied with the injector than his other duties. The probability is, therefore, that the engineer will run no such chance, and will operate his injector with the least risk of losing time, even though his method of feeding the boiler may not be the best for obtaining economy of fuel.

The temperature of the supply water does not often enter into the consideration, so long as the range of capacities of a locomotive injector is not affected by summer temperatures. Some of the South American and foreign railroads specify the limiting temperature of the water supply, but these cases are exceptional; the introduction of blow-back valves or the return of the air-brake exhaust to the tank has required the use of hot-water injectors in a few cases, but at the present time the use is not general, for it is generally accepted that no uncertainty should be allowed in the feed apparatus, and any increase in the temperature of the water supply renders the action of any injector less efficient and consequently reduces the length of life of the tubes.

REPAIRS.

Given an injector which has been accepted by the officials of a railroad as the most suitable, the next practical question which arises is that of maintenance and repair. If it fulfill the first-named requirement of simplicity of construction, it will not be difficult for the employee having special care of the boiler feeders to comprehend the ordinary cause of complaint on the part of the engineer or fireman who operates it. A few practical hints to the inspector may be of value.

The first duty on his part will be to examine the connections and joints to discover leaks or stoppage in the strainer, suction and delivery pipe, and the check valves. These proving to be in good condition, the injector should be tested under full working pressure. The methods then to follow depend to a great extent upon the style of injector, and its action during this test. Assuming it to be of the single-jet, fixed-nozzle class, like the Monitor, Ohio, Mack, etc., and that it cannot be made to prime, the fault probably lies with the lifting spindle or priming nozzles. If these nozzles are separate, as in the older form of Monitor, first examine the lifting steam nozzle to see if stopped up, then if tightly screwed to seat, and lastly, joint of body, as air may enter the combining tube at this place and so destroy the suction. The overflow nozzle may be bent, loose or out of line, or the drip pipe so close, or of such small size, that the free discharge of steam is impeded ; or the steam valve to the main steam nozzle may leak to such an extent as to fill the suction pipe and prevent the lifting of the supply water. With other patterns of injector in which the priming is effected by a central spindle discharging steam through the combining or delivery tube, such as the Sellers' 1876 Injector, the Ohio, Monitor 1888, etc., repeat the same method of procedure; examine the lifting jet, unscrewing it if necessary, to remove any obstructing particle of coal or scale. This lifting jet should be in exact line with the other tubes, and if bent, discharges the steam against the side of the other nozzle and the desired vacuum is not produced. If the injector is of the double-jet type, the Metropolitan, Hancock, Schutte, for instance, all the applicable tests above described should be used; also, ascertain if the position of the small spindle which regulates the steam pressure in the lifting nozzle permits a sufficient amount of steam to enter, and then if the steam valve to the forcer set is closed when the lifter steam valve is wide open; further, there may be interference with the free discharge of the steam through the body, owing to inaccurate setting of the relief or overflow valves due to wear of operating parts or faulty construction, or to the valves being prevented by some cause from lifting from their seats; in some cases, where loose valves or soft seats are used, they may have become separated from their attachments and obstructed some of the waste passages. In the Improved Sellers', 1887, and the 1897 Monitor, examine the inlet valve in the passage from the overflow chamber to the suction; this valve may not seat firmly, due to the bending of the stem or to some obstruction; this valve should be perfectly tight, or the discharged steam from the lifting nozzles can pass directly into the suction chamber.

If the preliminary test showed the lifting apparatus to be operative, and that the fault lay with the delivery of the feed to the boiler, the methods to be applied would depend upon the kind of injector and its action when steam was admitted to the forcing nozzles. If the injector be of the simpler type a heavy waste at overflow indicates a stoppage of the tubes, or that the delivery tube is worn too large for further use; the latter condition would be still more evident at the lower pressures. Another possible cause of trouble is the interference with the jet due to displacement, breakage of or obstruction by the line-check valve. In some forms of the double-jet type, the non-closing of the overflow relief valve between the lifter and the forcer sets prevents the proper inter-action of the two jets, and the feed water will not enter the boiler; or, the tubes of the lifter or forcer may be unequally worn, so that the correct ratios are lost, and either too little or too much water may be supplied to the forcer tubes.

The injector should then be removed from the locomotive, the parts carefully unscrewed and examined; if the diameters are worn too much, new parts should be substituted; exactly how much enlargement is permissible, depends upon the kind of service and the pattern of injector, some special styles permitting more latitude in this respect than others. One of the best practical tests of the superiority of an injector is the length of service without removal from the engine, and the cost of maintenance; the former may vary from one to six or seven years; the latter depends upon the design, the cost of repair parts, and the condition of the water and steam supplied to it. It is often possible to replace the old tubes if the diameters are not too much enlarged, after the interior surfaces have been carefully smoothed out with fine emery cloth, and all roughness and abrasion removed, for the slightest impediment to the rapid motion of the moving mass of water and steam interferes with the action of the jet.

All injectors should be tested before replacing upon the boiler. A simple form of testing plant,—such as that shown in Fig. 43,—should be part of the equipment of every large railroad shop conducting extensive injector repairs; it should be connected to a boiler capable of carrying as high pressure as that used on locomotives; the proving test should be made as complete as possible, and a permanent record preserved for reference.

FEEDING LOCOMOTIVE BOILERS.

In the economical handling of the locomotive, the way in which the boiler is fed plays an important part. To lay down a rigid rule is manifestly impossible, but that generally accepted, and adopted where conditions permit, is to maintain a constant water level with a continuous feed. There are a number of reasons for this: the strains on the boiler sheets, due to changes in temperature and unequal expansion, are reduced; the drain upon the steam supply is more constant, owing to the continuous operation of the injector; less water is wasted in stopping and starting the injector, and less time and attention need be given on the part of the operator; on long level runs, this method can almost always be pursued, if the engineer is provided with an injector having a wide range of capacities; but there are several special rules to be observed; in approaching a station at which a short stop is made, especially between long and fast runs, it is advantageous to stop the injector a short time before the station is reached, to permit a slight checking of the fire, and then, when the station is reached, to feed the boilers quickly with one, or even with both injectors if necessary, to prevent blowing off at the safety-valve. Also, the feed should be stopped before starting from a station; the train can then be drawn out with full boiler pressure, and there need be no drain on the steam supply due to use of the injector, until the train is under good headway and the exhaust has pulled the fire into good working condition. With a badly steaming or overloaded engine drawing a heavy train it is very difficult to raise the pressure to the normal, having a low water level to overcome in addition.

A curious phenomenon, often observed in connection with the feeding of locomotive boilers, is the variation in the water level after an injector has been started; when a boiler has not fed for some time and the engine has been working hard with a heavy exhaust, clear fire and waste gases at a high temperature, the entire body of water is filled with bubbles of steam working upward to the surface; this mass of confined air and steam displaces a large volume of water, while the agitation of the surface due to the motion of the engine tends still further to raise the apparent water level. If the boiler feeder is now started, the cooler water condenses the rising steam bubbles and diminishes the volume of water, and lowers the level in the glass. In

FEEDING LOCOMOTIVE BOILERS. 157

former days, when pumps were used, five to ten minutes were often required before any increase in the height of water would be shown, and although the injector would not produce so marked an effect, yet a deceptive contraction in the volume does occur, which should be carefully watched; otherwise, when normal conditions of ebullition are established, it will be found that the water level has risen very suddenly, at the probable risk of water passing over into the cylinder.

In order to obtain the most economical results, every device used on and about a locomotive should be of the most efficient pattern. No operator, whether on a railroad or elsewhere, can do his best for his employer unless he is supplied with suitable appliances or tools, and given the means whereby he can be most economical of his own effort and the material at his disposal. This is especially true of the injector, which should be arranged to give good results with the fewest movements and with the least inconvenience to the operator. This is an important and practical question: if an engineer has several handles to manipulate when one should answer as well, or if fine and careful adjustment is required to give good results, or if there be risk of breakage of the jet when throttling to the minimum, the engineer will operate his injector as best suits his convenience, with little reference to the finer points of economy. On the other hand, all good men are ambitious and anxious to do their best and make records for themselves which may tend to their advancement; if an engineer or fireman feels that he can obtain good results and will receive material encouragement in his efforts, he will steadily improve. Everything about a locomotive should be the best of its kind, whether air-brake, lubricator or boiler-feeder, and should be carefully tested by the motive-power department, and full evidence obtained that it approaches most nearly the standard of modern excellence. Great advancement has already been made, yet further improvements are expected at the hands of those who are working toward the improvement of the injector, along the lines laid down at the opening of this chapter.

In conclusion, a word might be said regarding the careful handling of injectors, which are not delicate instruments, yet require occasional attention in order to give the best satisfaction. A little care in closing the valves and pushing in levers will prolong the life of the parts and reduce the necessity for grinding and repairing. If dirt or scale is caught between a valve and its seat, causing a slight leak, a touch with a seat reamer or a little time taken to grind the surfaces to a bearing, will save much future trouble. All valves should be kept tight, as a leak tends invariably to increase rapidly, and the trouble is magnified if repairs are postponed. The deposit of a hard scale upon the interior surfaces of the tubes and in the overflow vents is a continual source of trouble when the feed water contains lime in solution; a small amount of oil fed into the steam or suction pipes alleviates but does not entirely prevent the formation of the deposit; it may, however, in most cases be removed by allowing the tubes to remain for eight or ten hours in solution of one part of muriatic acid in ten of water.

The repairing of injectors should never be intrusted to a novice, and where large numbers are in use, it is customary on large railroads to combine the air brake and injector repairs in the same department, and place the responsibility in the hands of experienced workmen; every opportunity should be afforded these men for obtaining a thorough knowledge of the theory of action, and the special peculiarity of each pattern of injector in use, and it has been the experience of the author that courteous attention is always given by the well-known manufacturing companies to those who seek for information.

INDEX.

AIR IN STEAM, 72.
AUTOMATIC DOUBLE JET, 23.
AUTOMATIC INJECTOR—Definition of, 27; Earliest Forms of, 22; Exhaust, 113; Gresham, 110; Manhattan, 111; Metropolitan, 109; Penberthy, 110-119; Sellers' 1885, 108; Sellers' 1887, 104; Webb, 92.
BACK PRESSURE, 119; Effect of Feed-Temperature Table, 72; Experiments, 34, 122.
BANCROFT, J. S.—Improvements by, 22
BELFIELD INJECTOR, 66; Description, 102.
BOURDON—Inventions by, 3.
BOILER EVAPORATION, 126.
BOILER TESTERS, 121.
BUFFALO INJECTOR, 120.
CAPACITY—See Weight of Water per Pound of Steam.
COEFFICIENT OF IMPACT, 77, 78.
COMBINING TUBE, 45, 69; Definition of, 26; Losses in, 74; Vacuum Within, 49; Wear of, 51.
DELIVERY TEMPERATURE, 72, 78, 106.
DELIVERY TUBE—Definition of, 26; Pressure in, 33; Wear of, 42.
DENSITY OF JET—Effect of Temperature of Feed, 78; in Delivery Tube, 31, 72.
DESMOND—Improvements by, 22.
DIAGRAM—Maximum and Minimum Capacities at different Feed Temperatures, 127; Maximum and Minimum Capacities at different Pressures, 125, 144; Pressures in Delivery Tube, 40; Pressure within Steam Nozzle, 57, 60; Velocity and Pressure in Delivery Tube, 32.
DIVERGENT FLARE—Delivery Tube, 34, 39; Steam Nozzle, 54.
DIVERGENT STEAM NOZZLE—Invention of, 21.
DOUBLE JET—Automatic Form of, 23; Definition of, 27; Effect of Height of Lift, 77; Pressure between the two sets of tubes, 78; Relative Proportions to Single Jet, 78.
DOUBLE JET INJECTOR—Belfield, 102; Description of, 19; for High Feed Temperature, 50; Schutte, 101.
EBERMAN INJECTOR, 114.
EFFICIENCY—Cause of loss of, 86; Compared with Pump, 87; Definition of, 28; Mechanical, 84, 86; of various types, 66.
EJECTORS, 123; Definition of, 25.
ENGLISH INJECTORS, 90.
ENGLAND — Introduction of Injector into, 4.
EXHAUST INJECTOR, 64; Description of, 118; Length of Combining Tube, 48; Theory of Action, 70.
FEED WATER—Purity of, 51; Temperature of, 50, 101, 106, 114, 119, 147.

FEED TEMPERATURE—Effect upon Capacity, 78; Effect on Maximum and Minimum Capacity (Diagram), 133; Highest Admissible (Formula for), 85.
FEEDING LOCOMOTIVE BOILERS, 155.
FIRE PUMPS, 123.
FRANCE—Improvement in, 20.
FRENCH INJECTORS, 93.
FRIEDMAN—Improvements by, 21.
GARFIELD INJECTOR, 32, 66, 109.
GIFFARD, H. J., Original inventor, 1.
GIFFARD INJECTOR, 10-32.
GRESHAM AND CRAVEN, 92.
GRESHAM, JAMES—Improvements by, 18.
GRESHAM RE-STARTING INJECTOR— Description of, 116.
HAMER, METCALF AND DAVIES' EXHAUST INJECTOR, 21.
HANCOCK INJECTOR, 109.
HANCOCK, JOHN—Improvements by, 18.
HOLDEN AND BROOKE—Improvements by, 23.
HORSE POWER—Required to Feed Boiler, 65.
IMPACT—Coefficient of (Table), 72; Losses during, 86.
INJECTOR—Definition of Term, 25.
IRWIN INJECTOR, 48.
ITALIAN INJECTORS, 93.
JENKS INJECTOR, 120.
KNEASS, STRICKLAND L.—Improvements by, 23.
KORTING, ERNEST—Improvements by, 18.
LIFT—Effect upon Capacity (Table), 76.
LITTLE GIANT INJECTOR (RUE), 66; Description of, 113; Experiments with, 46.
LOCOMOTIVE INJECTORS—Sizes Required (Table), 121.
LOFTUS, JOHN—Improvements by, 22.
LOSSES BY CONDUCTION OF HEAT, 83.

LOSSES IN COMBINING TUBE, 69, 74.
MACK INJECTOR, 32, 66, 107.
MANHATTAN INJECTOR—Description of, 117.
MAXIMUM CAPACITY—Conditions affecting, 119.
MECHANICAL EFFICIENCY, 64.
METROPOLITAN INJECTOR, 32, 66, 108, 121.
MILLHOLLAND—Improvements by, 13.
MINIMUM CAPACITY—(Formula), 84.
MONITOR INJECTOR, 32, 36; Description of, 97.
MONITOR STANDARD, 90.
NATHAN MFG. CO., 94.
NON-LIFTING INJECTOR, 89.
OHIO INJECTOR, 109.
OIL—Use of to Prevent Scale, 129.
PARK INJECTOR, 120.
PENBERTHY, WILLIAM—Improvements by, 22.
PENBERTHY INJECTOR—Description of, 118; "SPECIAL," 110.
PENNA. R. R. STANDARD, 90, 129.
PRESSURES—Within Delivery Tube (Diagram of), 32; Within Steam Nozzle, 57.
PUMP—Compared with Injector, 87.
RANGE OF CAPACITY—Definition of, 28.
RE-STARTING—Sellers', 111.
REPAIRS, 153.
ROBINSON AND GRESHAM—Improvements by, 13.
RUE INJECTOR—See "LITTLE GIANT."
RUE, SAMUEL—Improvements by, 13.
SCALE—Prevention of, 158.
SCHAU—Improvements by, 21.
SCHUTTE (KORTING) INJECTOR—Description of, 100.
SELF-ADJUSTING INJECTOR—Action of Combining Tube, 48; Advantages of, 47; Definition of, 27; Description of, 15, 96.
SELLERS, WM.—Inventor of Self-Adjusting Principle, 14.
SELLERS' 1876 INJECTOR, 32.

INDEX. 161

SELLERS' 1885 INJECTOR—Description of, 108; High Feed Temperature, 51.
SELLERS' RE-STARTING, 111.
SELLERS' 1887 INJECTOR, 66; Description of, 103.
SETTING INJECTORS, 120.
STEAM DISCHARGE, 58; Phenomena of, 54; Rate of, 61, 63; Formulæ, 61, 63.
STEAM JETS—Air mixed with, 71.
STEAM NOZZLE— Definition of, 26; Ratios Throat to Initial Pressures, 58, 60, 61.
STEAM PRESSURE — Effect on Maximum and Minimum Capacity (Diagram), 124, 144.
STEWART, ROBINSON AND GRESHAM— Improvements by, 20.
TABLE I.—Delivery of various Injectors per Unit Area of Delivery Tube, 32.
TABLE II.—Pressure and Weight of Water, 43.
TABLE III.—Values of θ, 65.
TABLE IV —Density of Jet in Delivery Tube, 72.
TABLE V.—Variation of Weight of Water per Pound of Steam with Different Heights of Lift, 76.
TABLE VI. — Tension of Vapor of Water, 77.
TABLE VII.—Variation of Capacity with Feed Temperature, 78.

TABLE VIII.—Weight of Water per Pound of Steam at Different Steam Pressures, 79.
TESTS OF INJECTORS, 129.
THEORY OF INJECTOR, 67.
VABE—Early Form of Automatic Injector, 22.
VARIATION OF DENSITY OF JET WITH FEED TEMPERATURE (Table), 72, 78.
VELOCITY OF STEAM DISCHARGE, 55; At various pressures, 61; Formulæ, 61; At Impact, 72.
WATER—Weight of 1 Cubic Foot of, at various Temperatures, 43.
WATER ENTRANCE TO COMBINING TUBE, 46.
WATER IN STEAM, 64.
WEAR—Combining Tube, 51.
WEBB'S INJECTOR, 93.
W. F. INJECTOR — Description of, 99.
WEIGHT OF STEAM DISCHARGE, 61; Experiments and Formulæ, 63.
WEIGHT OF WATER—Delivered per Sq. Mm. of Delivery Tube, 66; per Pound of Steam (Formula for), 76, 83; per Pound of Steam at different Steam Pressures, 79; per Pound of Steam, 72, 146; Variation with Feed Temperature, 78.
WILLIAMS, G. C.—Automatic Injector, 22.

www.ingramcontent.com/pod-product-compliance
Lightning Source LLC
Chambersburg PA
CBHW030246170426
43202CB00009B/640